EARTH: GEOLOGIC PRINCIPLES AND HISTORY

EARTH: GEOLOGIC PRINCIPLES AND HISTORY

Stanley Chernicoff
University of Seattle, Washington

Haydn A. "Chip" Fox
Texas A & M University, Commerce

Lawrence H. Tanner
Bloomsburg University

Houghton Mifflin Company Boston New York

Editor-in-Chief: Kathi Prancan
Development Editor: Marianne Stepanian
Senior Project Editor: Cathy Labresh Brooks
Editorial Assistant: Rachel Levison
Senior Production/Design Coordinator: Jill Haber
Manufacturing Manager: Florence Cadran
Executive Marketing Manager: Andy Fisher

Cover photos: © 1997 Artville, LLC.

Photo credits begin on page C-1.

Printed in the U.S.A.

Library of Congress Control Number: 2001090880

ISBN: 0-618-02275-9

1 2 3 4 5 6 7 8 9-WC-05 04 03 02 01

To my wife, Dr. Julie Stein, and my sons, Matthew and David, who have tolerated my obsession with text writing with extraordinary grace and humor (and endured an uncountable number of interrupted family dinners). Much thanks and appreciation for your remarkable support and encouragement.

Stan Chernicoff

To my wife, Jannie B. Fox, whose patience and encouragement were of limitless value in the writing of this book and in life in general.

Chip Fox

To Dorothy and Edmund, who nurtured my early interest in the sciences in so many ways; to Linda, for giving me the courage to follow my geologic passions to any destination; and to Emily, that she also may find a road to follow that beckons as strongly; this book is affectionately dedicated.

Lawrence H. Tanner

About the Authors

Born in Brooklyn, New York, Stan Chernicoff began his academic career as a political science major at Brooklyn College of the City University of New York. On graduation, he intended to enter law school and pursue a career in constitutional law. He had, however, the good fortune to take geology as his last requirement for graduation in the spring of his senior year, and he was so thoroughly captivated by it that his plans were forever changed.

After an intensive post-baccalaureate program of physics, calculus, chemistry, and geology, Stan entered the University of Minnesota–Twin Cities, where he received his doctorate in Glacial and Quaternary Geology under the guidance of one of North America's preeminent glacial geologists, Dr. H. E. Wright. Stan launched his career as a purveyor of geological knowledge as a senior graduate student teaching physical geology to hundreds of bright Minnesotans.

Stan has been a member of the faculty of the Department of Geological Sciences at the University of Washington in Seattle since 1981, where he has won several teaching awards. At Washington, he has taught Physical Geology, the Great Ice Ages, and the Geology of the Pacific Northwest to more than 20,000 students, and he has trained hundreds of graduate teaching assistants in the art of bringing geology alive for nonscience majors. Stan studies the glacial history of the Puget Sound region and pursues his true passion, coaching his sons and their buddies in soccer, baseball, and basketball. He lives in Seattle with his wife, Dr. Julie Stein, a professor of archaeology, and their two sons, Matthew (the midfielder, second baseman, two-guard) and David (the striker, second baseman, point guard).

Born in Grand Rapids, Michigan, Haydn A. "Chip" Fox received a bachelor's degree in theology and journalism from Ambassador College in Big Sandy, Texas, in 1971. He spent the next 15 years in several nonacademic pursuits, such as managing convenience stores, driving a truck, and working for a newspaper. In 1986, he returned to college with the intent of becoming an earth science teacher. His interests in geology, earth science, environmental science, and science education were immediately sparked, and in 1992 he received a Ph.D. in Geological Sciences from the University of South Carolina.

Chip has served on the faculty of departments of Earth Science at Southeast Missouri State University and Clemson University. He is currently a member of the faculty at Texas A & M University in Commerce, Texas, where he teaches numerous courses in earth science and the environmental sciences. Having spent several years in nonacademic careers, Chip understands the challenges his students will face when they graduate from college. He enjoys an excellent rapport with his students, a group that includes several future geologists, environmental professionals, and public school teachers.

Chip lives with his wife, Jannie, in an old farmhouse in Texas, where they are busy remodeling and trying to make a tenuous paradise of their 54 acres.

About the Authors

Born in Niskayuna, New York, Larry Tanner was raised in Pittsfield, in western Massachusetts, an area largely devoid of fossils. Fortunately, the collection at the local museum and the abundance of fossils in the backyard of his grandmother's house, near Ithaca, New York, made up for this misfortune and fueled an early interest in geology. Tanner attended Williams College in Williamstown, Massachusetts, initially as a biology major; however, a series of labs with fruit flies eventually convinced him to renew his interest in his childhood hobby of geology. Following graduation, Larry continued his study of geology at Tulsa University and worked for a number of years as a production geologist in the petroleum industry. Not satisfied with the extent of his geologic education, Larry left industry and returned to graduate study at the University of Massachusetts at Amherst. Here he pursued a doctorate in sedimentary geology under the direction of Professor John Hubert, who introduced him to the mysteries of the Mesozoic world.

Larry came to Bloomsburg University of Pennsylvania in 1992 following teaching stints at Colby College and Colgate University. A professor of geology, Larry teaches Physical Geology, Historical Geology, Natural Disasters, Earth Materials, and Sedimentation and Stratigraphy. His undergraduate students have accompanied and assisted him in his field research in such locations as Hawaii, Death Valley, the Four Corners of the Southwest, and the Canadian Maritimes. Among all of his research areas, however, unraveling the paleoclimate and paleogeography of the Mesozoic remains his primary interest. Larry lives in Winfield, Pennsylvania, with his wife, Linda LeMura, a professor of exercise physiology; their accomplished pianist/soccer star daughter Emily; and Checkers the cat.

Contents in Brief

Contents

2 Minerals

3 Igneous Processes and Igneous Rocks

Preface

The introductory geology courses, taken predominantly by nonscience majors, may be the only science courses taken by some students during their college years. What a wonderful opportunity this provides us to introduce students to the field we love and show them how fascinating the Earth is. Indeed, much of what students will learn in Physical and Historical Geology will be recalled throughout their lives, as they travel across this and other continents, sit on beaches, dig in their gardens, and visit our national parks. For this reason, our book team—authors, illustrators, photo researchers, and editors—has expended its best efforts to craft an exciting, stimulating, and enduring introduction to the field.

The Book's Goal

This book's goal is basic—to teach what everyone should know about the Earth in a manner that will engage and stimulate. It embodies the view that geology is perhaps the most useful college-level science subject a nonscience major can study—and we truly believe it is one all students should take. Geology can show students the essence of the scientific process and how scientists work, and at the same time nurture an appreciation and respect for their natural environment by creating an understanding of its history, the physical processes that shape it, and the natural resources and hazards resulting from these processes. It is through the study of geology that students best can be prepared to take their place in a society that must cope with threats to the environment and responsibly utilize the resources of the Earth to provide for the wellbeing of all.

Contents and Organization

Earth: Geologic Principles and History presents the most important concepts of physical and historical geology in a style designed for nonmajors. The result is a text embodying all of the major themes and ideas about the Earth—its composition, principal processes, and history—presented in a clear and concise format that avoids detailed discussion of the implications of and alternatives to the major theories of geology. Even while intended for nonmajors, this text is certain to excite and stimulate some students into considering geology as a field of study within their interests and capabilities.

In Chapter 1, initial discussion of the origin of the Earth, the rock cycle, and geologic time builds a foundation for the succeeding chapters. Chapters 1 through 8 introduce the basics—minerals, rocks, and time. Part 2 of the book, Chapters 9 through 11, discusses structural geology, earthquakes, the Earth's interior, and the details of plate tectonics. Part 3, Chapters 12 through 18, presents the principal geomorphic processes of mass movement, streams, groundwater, glaciers, and desert processes. Important unifying themes that were introduced in Chapter 1, such as plate tectonics, environmental geology, and natural resources, are revisited in their proper context in later chapters throughout these first three sections. Part 4 of the book, Chapters 19 through 26, summarizes our knowledge of the history of physical and biological change on our planet. Following an introduction to the study of historical geology in Chapter 19, the major geologic and evolutionary developments are described in a clear chronological format.

Artwork

The drawings in this book are unique. As you will see when you leaf through this book, the art explains, describes, stimulates, and teaches. It is not schematic; it shows how the Earth and its geological features actually look. It is also not static; it shows geological processes in action, allowing students to see how geological features evolve through time. Every effort has been made to illustrate accurately a wide range of geological and geomorphic settings, including vegetation and wildlife, weathering patterns, even the shadows cast by the Sun at various latitudes. The artistic style is consistent throughout, so that students may become familiar with the appearance of some features even before reading about them in subsequent chapters. For example, the stream drainage patterns appearing on volcanoes in Chapter 4, Volcanoes and Volcanism, set the stage for the discussion of drainage patterns in Chapter 13, Streams and Floods. The colors used and the map symbols keyed to various rock types follow international conventions and are consistent throughout.

Pedagogy

Nearly every chapter begins with a particularly fascinating aspect of the subject of the chapter, designed to captivate the student's interest. Most chapters also contain one or more Highlights—in-depth discussions of topics of popular interest. In many instances, the Highlights comprise late-breaking stories that demonstrate the relevance and impact of geology on our lives, or alternatively, how new scientific discoveries change our ideas about established theories.

To help readers learn and retain the important principles, every chapter ends with a Summary, a narrative discussion that recalls all of the important chapter concepts. Key terms, which are in boldface type in the chapter, are listed at the chapter's end and also appear in boldface in the Summary. Also at the end of every chapter are two question sets: *Questions for Review* helps students retain the facts presented, and *For Further Thought* challenges readers to think more deeply about the implications of the material studied.

The authors and illustrators have tried to introduce readers to world geology. This book emphasizes, however, the geology of North America (including the offshore state, Hawai'i), while acknowledging that geological processes do not stop at national boundaries or at the continent's coasts. Wherever data are available—from the distribution of coal to the survey of seismic hazards—we have tried to show our readers as much of this continent, and beyond, as feasible. Photos and examples have been selected from throughout the United States and Canada and from many other regions of the world.

The metric system is used for all numerical units, with their English equivalents in parentheses, so that U.S. students can become more familiar with the units of measurement used by virtually every other country in the world.

Supplements Package

Earth: Geologic Principles and History is accompanied by an array of materials to enhance teaching and learning.

Students who wish additional help mastering the text can use the **Study Guide** by W. Carl Shellenberger (Montana State University—Northern), which also accompanies Stanley Chernicoff's *Geology,* Second Edition as well as *Essentials of Geology,* Second Edition. For each chapter, the Guided Study section helps students focus on and review in writing the key ideas of each section of the chapter as they read. The Chapter Review, arranged by section and composed of fill-in statements, enables them to see if they have retained the ideas and terminology introduced in the chapter. The Practice Tests and the Challenge Tests, which consist of multiple-choice, true/false, and brief essay questions, test their mastery of the material. In addition, study guide material will be available on the Houghton Mifflin geology web site for the eight historical chapters (Chapters 19–26) presented in *Earth: Geologic Principles and History.*

The **Instructor's Resource Manual** by Chip Fox accompanying *Geology,* Second Edition and *Essentials of Geology,* Second Edition features an outline lecture guide with teaching suggestions embedded in it and student activities and classroom demonstrations. Answers to the end-of-chapter questions in the textbook are also provided. Also included is a comprehensive **Test Bank** that contains more than one thousand questions. There are at least 40 multiple-choice questions per chapter, classified either as factual or conceptual/analytical. There are also ten short essay questions, complete with answers for each chapter. A computerized version of the Test Bank is available in both IBM and Macintosh formats. In addition, instructor's resource material and test questions keyed to the eight historical chapters (Chapters 19–26) will be available on the Houghton Mifflin geology web site.

Also available is the **Geology Laboratory Manual** by James D. Meyers, James E. McClurg, and Charles L. Angevine of the University of Wyoming. This inexpensive manual is closely tied to the text and offers twenty physical geology labs on topics such as maps, plate tectonics, sedimentary and metamorphic rocks, streams, and groundwater. Each lab contains multiple activities to develop and hone students' geological skills. Worksheets are designed to be torn from the manual and submitted for grading.

An extensive, full-color **transparency** package containing the text's diagrams is available for classroom use. Select photos from the text will be available in PowerPoint format for download from the text web site.

The book is supported further by its award-winning web site, GEOLOGYLINK (found at www.geologylink.com), maintained by its web masters, Dr. Robert Nelson and Laura Jean Wilcox, from Colby College, Waterville, Maine. This site will tell you what of geological import has happened overnight while you slept. It also contains expanded discussions of "hot topics" in the field of geology and an exhaustive encyclopedia of links to all things geological. For the first edition of *Earth: Geologic Principles and History,* our web site contains chapter quizzes, tutorials, electronic flashcards, as well as on-line version of the *Peterson's Field Guide to Rocks and Minerals* by Frederick Pough. These outstanding teaching and learning aids help the student learn the principles of physical and historical geology through multimedia technology and enable them to master geologic principles in a stimulating and thoughtful environment.

Acknowledgments

Richard Bonnett, Marshall University

Patricia J. Bush, Delgado Community College

Paul J. Bybee, Utah Valley State College

Robert M. Chandler, Georgia College and State University

Larry Coats, Northern Arizona University

Emily Cobabe, Montana State University

David Gibson, University of Maine at Farmington

Laura A. Guertin, University of Colorado

Lance E. Kearns, James Madison University

David King, Auburn University

Steve May, Walla Walla Community College

William N. Mode, University of Wisconsin-Oshkosh

Shannon O'Dunn, Grossmont College

William Orr, University of Oregon

Keith Putirka, Indiana University of Pennsylvania

Robert W. Reynolds, Central Oregon Community College

Bethany D. Rinard, Tarleton State University

Kelvin Rodolfo, University of Illinois

Glenn Stracher, East Georgia College

Jody Tinsley, Clemson University

Mark A. Wilson, The College of Wooster

Thomas Yancey, Texas A&M University

Grant M. Young, University of Western Ontario

Carol Ziegler, University of San Diego

We also give special thanks to the people of Houghton Mifflin who guided this text through its various stages of development and production: Kathi Prancan, Editor-in-Chief; Marianne Stepanian, Development Editor; Karen O'Connor, Editorial Associate; Cathy Brooks, Senior Project Editor; Jill Haber, Senior Production/Design Coordinator; Charlotte Miller, Art Editor; Naomi Kornhauser, Photo Editor; Andy Fisher, Executive Marketing Manager.

Thanks also to the artists at J/B Woolsey Associates for their fine illustrations: John D. Woolsey, Art Developer and Editor; Craig Durant, Art Director; and Artists Paul Palcko, Kelly Paralis, Kandis Elliot, and Rick Jones.

We have used our teaching experiences to craft a textbook that we think our own students will learn from and enjoy. We hope your students will, too. We invite your comments: please send them to the authors, whose email addresses are: sechern@u.washington.edu, haydn_fox@tamu-commerce.edu, lhtann@bloomu.edu.

To the Student

Why study geology? To learn something about the planet on which we live, for geology is the scientific study of almost all physical aspects of the Earth. We see geology all around us every day; hills, streams, oceans—all of these are formed by geologic processes that have worked to shape the Earth for billions of years. These processes give us the water and soil on which we depend for food, produce most of the energy used by our society, and provide the mineral resources on which our industries depend. Our planet is very dynamic. Geology can be exciting as we witness volcanic eruptions or view mountainous landscapes; it is also tragic when natural disasters, such as earthquakes or floods, destroy homes and lives. Studying geology gives us the power that comes with understanding these processes, power that allows to better protect ourselves from tragedy and utilize the resources of the Earth. Our planet also has a long and fascinating history during which its appearance—the size and positions of oceans and continents—has changed tremendously. The story of these physical changes, and of the organisms that evolved to inhabit the Earth, is more remarkable than any fiction.

There were only a few dozen dinosaur genera known when I was six years old, and I knew the names of most of them. But in the years since, this number has increased roughly ten-fold, and most other areas of scientific knowledge have increased similarly. For this reason, this book can present only a brief survey of the incredibly broad realm of geologic knowledge, ranging from the composition of the Earth, to the processes that shape our planet, and including its history from its origin to the present. Don't be intimidated by the volume of information contained in this text. I am confident that you will find particular aspects of this survey that you can understand simply by looking at the geology of your home.

I began my undergraduate education with every intention of becoming a research biologist, but somewhere along the way I realized that spending days in a laboratory wearing a white coat was not for me. Rather late in my college career I enrolled in an introductory geology course and became hooked. Every hill, every valley, each rock could be explained as the result of geologic processes. The theory of plate tectonics was still quite new then and it allowed us to understand the formation of mountains and oceans, the eruptions of volcanoes, and the causes of earthquakes. Best of all, here was a field of science that actually required me to spend lots of time outside studying the landscape, examining rock formations, and looking for fossils.

Part of the appeal of geology for many of us is indeed the fact that, because it is the study of geologic processes and products, we must spend a great deal of time in the field. This is the best classroom of all, but the type of instruction you will receive in this setting cannot be found in textbooks. Here you must depend on your professors who will be your guides, teachers, and friends. I was fortunate to have some outstanding mentors throughout my student career. From Bill Fox, one of my professors at Williams College, I first gained an understanding that the sedimentary processes produce very predictable products. Years later, this appreciation was deepened by John Hubert, my mentor at the University of Massachusetts, who first led me into Mesozoic landscapes and taught me how geologists reconstruct the ancient world by playing detective. If you are lucky, you too will have professors who will share their wisdom and enthusiasm for the study of the Earth.

Lawrence H. Tanner

Features of

Earth: Geologic Principles and History

The rich detail and technical accuracy of the illustrations help convey complex concepts to introductory students.

Photos are often paired with art to emphasize a point.

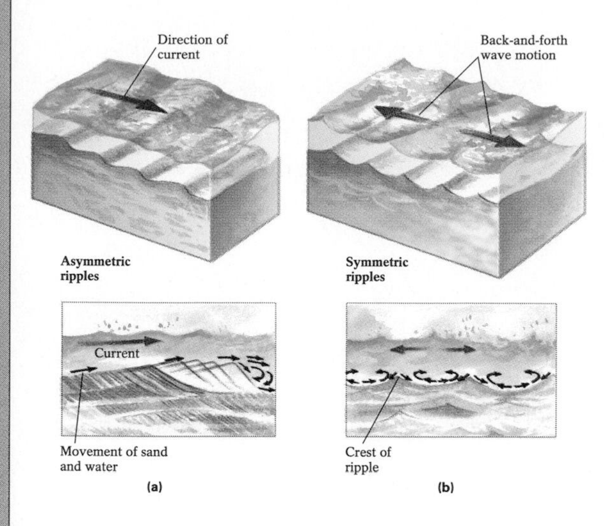

Figure 6-7 Different types of currents produce different ripple patterns. **(a)** A current that generally flows in one direction, such as a stream, produces asymmetric ripples. Sand grains roll up the gently sloping upstream side of each ridge and then cascade down the steeper downstream side. **(b)** Symmetric ripples form from the back-and-forth motion of waves in shallow surf zones at the coast or at the water's edge in a lake. Photo: Exposed rocks show ripple marks, evidence of past current flow, either water or wind.

series of shallow curving ridges. The configuration of these ridges, which are often visible on sandy surfaces, reflects the nature of the current that produced them (Fig. 6-7).

Mudcracks are fractures that develop when the surface of wet fine-grained sediment (mud) dries and contracts (Fig. 6-8). Because these structures form only at the top of a layer of muddy sediment and narrow progressively downward, geologists can study mudcracks to determine whether a layer of sedimentary rock has been overturned.

Lithification: Turning Sediment into Sedimentary Rock

When a sediment layer is deposited, it buries all previous layers deposited at that location. Eventually, the continuing deposition may enable a sedimentary pile to become several kilometers deep. Such deep burial may convert sediments into solid sedimentary rock by the process of **lithification** (from the Greek *lithos,* meaning "rock," and Latin *facere,* meaning "to make"). During lithification, sediment grains become compacted, often cemented, and sometimes recrystallized.

Compaction is the process by which the volume of buried sediment, either detrital or chemical, becomes diminished by pressure exerted by the weight of overlying sediments. Expulsion of air and water from the sediment and the reduction of the spaces between grains combine to

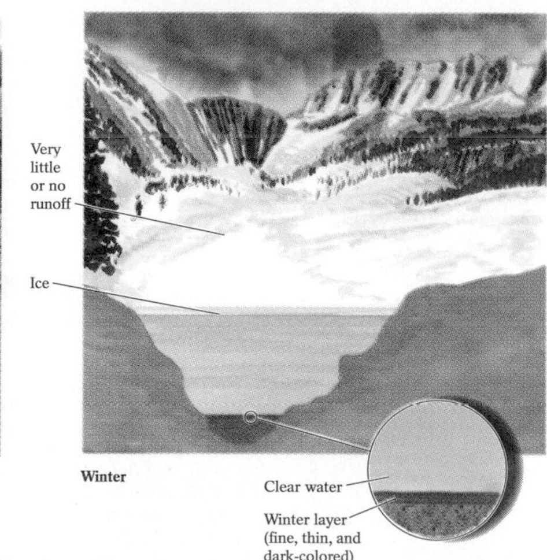

Figure 8-13 The origin of lake varves. A typical varve includes a thick, coarse, light-colored summertime layer produced during high runoff (from snowmelt and spring storms) and high sediment influx, plus a thin, fine dark-colored wintertime layer produced during low runoff and low sediment influx (or no influx, if the lake is frozen). Note the varves in the photo. Why do you think the varves vary so noticeably in thickness?

Figure 8-14 Lichen colonies on a granite boulder. The light-colored areas of rock have been bleached by chemicals in the lichen. The sizes of such colonies can provide clues as to how long the rock surface has been exposed.

deposited in summer and a thin, fine, dark-colored layer deposited in winter. By applying *varve chronology,* in which they study the number and nature of the varves underlying a lake, geologists can determine how long ago the lake formed and identify events, such as landslides, that affected sedimentation in the area (Fig. 8-13).

Lichen (pronounced "LIE-ken"), colonies of simple, plant-like organisms that grow on exposed rock surfaces, are the basis of a dating method known as *lichenometry*. Lichen grow extremely slowly; given similar rocks and climatic conditions, the larger the lichen colony, the longer the period of time since the growth surface was exposed. Study of these organisms can yield accurate dates for young glacial deposits, rockfalls, and mudflows—all events that expose new rock surfaces on which lichen can grow (Fig. 8-14).

Artwork shows geologic features in a naturalistic context, to give students a sense of how these features actually look.

Highlight 12-2 How to Choose a Stable Home Site

If you have not already done so, you may eventually wish to purchase your own home. Having studied physical geology, you will want to ensure that your dream house doesn't fall victim to a geological nightmare.

Suppose you're exploring southern California's scenic beachfront locales. You happen upon the mosaic sign for the town of Portuguese Bend and notice that the beautiful ceramic signpost is cracked into two pieces. You check the local real estate listings and find a house priced at $50,000 that should be worth $500,000. Your knowledge of geology, along with your common sense, immediately warns you that something may be wrong here. What else should you look for? How can you tell—in southern California or anywhere else in the world—if you're in mass-movement country?

Examine both the property itself and the entire neighborhood. Do you see any signs of an old mud or debris flow? Is there evidence of a slump scar upslope where a block may have broken away? As you drive through the community and its environs, look for fences that are out of alignment, and for power and telephone lines that seem too slack in some places and too taut in others (Fig. 1).

Next, look carefully at the house itself and, if possible, at neighboring ones. Search for large cracks in the foundation (small cracks may be due to initial drying and settling of the concrete). Doors and windows that stick may indicate that once-linear structural features are now out of line, although poor craftsmanship or high moisture content may also be responsible. A cracked pool lining might explain why a swimming pool doesn't retain water. Finding only one such problem may not indicate danger, but the presence of several problems should send a strong warning signal. If the geology, topography, and hydrology of a home site all raise questions about slope stability, the site may well be prone to progressive slope failure. To confirm your suspicions, try checking newspaper accounts and the records of the local housing authority, contacting the state geological survey, and interviewing the property's neighbors.

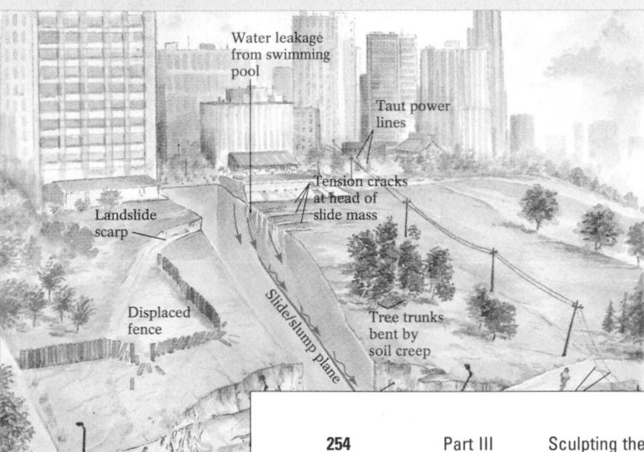

Figure 1 Various signs of past, current, and potential future mass movement in an urban area. If the power lines in a neighborhood are very loose, it suggests that the poles that hold them have moved closer together, as one might expect at the toe of a slide where the slope is bunching up like a rumpled carpet. At the head of the slide, taut lines may indicate that the poles on the slide mass are moving away from those upslope. The poles in the middle of a slide mass may keep their original spacing, because the mass may not be deforming much internally.

Highlight boxes introduce high-interest topics, making geology relevant to students' lives.

Highlight 13-2 Did We Help Cause the Midwestern Flood of 1993?

In the beginning of this chapter, we discussed one of North America's worst floods, which took place along the upper portions of the Mississippi River in the summer of 1993 (Fig. 1). Did human activity—65 years of "managing" the Mississippi River to prevent floods—actually contribute to the magnitude and tragedy of the flood?

Before the last 100 years or so, the Mississippi and its tributaries determined their own boundaries. During periods of high water, the rivers broke through or overflowed natural levees, flooding tens of thousands of square kilometers of the surrounding, largely uninhabited, lands. In the twentieth century, however, millions of people migrated to the region, building cities, towns, and large farms along the rivers' banks. When a flood in 1927 took 214 lives, Congress enacted the first Mississippi River Flood Control Act, assigning the U.S. Army Corps of Engineers the daunting task of confining the river to its channel.

The Corps of Engineers' efforts consist of about 300 dams and reservoirs and thousands of kilometers of artificial levees and concrete flood walls, all designed to prevent the river from spilling onto its natural floodplain. The system also contains numerous pumping stations, spillways, and diversion channels designed to divert water for storage in temporary holding basins. But the events of 1993 showed that such structures simply cannot contain an extraordinary flood. By confining such great discharges to a channel, the artificial retention structures actually caused the swollen rivers to flow more rapidly and violently, thus damaging the very structures designed to restrain them. By denying the river access to its natural floodplains, the structures caused the streams to rise higher than they would have otherwise, ensuring that once they did breach the levees, the floods would cause greater damage. Furthermore, the existence of artificial levees and flood walls had encouraged the growth of cities, towns, and farms closer to the riverbanks than was really safe.

What does the future hold for the residents of the upper Mississippi valley? Certainly more flooding, but perhaps less human interference with the river's natural behavior. Some communities have proposed that all flood-retention systems be eliminated and that zoning limit future development within the river's floodplain. Others look longingly at St. Louis's 16-meter (52-foot)-high concrete flood wall, which saved that city's downtown business district when the Mississippi reached its record crest at 14.2 meters (47 feet). The debate continues between those who believe we can tame the mighty Mississippi and those who believe we cannot.

Figure 1 Flooding at Kaskaskia, Illinois.

Chapter Summary

Volcanism, the set of processes that results in extrusion of molten rock, begins with the creation of magma by the melting of preexisting rock and culminates with the ascent of this magma to the Earth's surface through fractures, faults, and other cracks in the lithosphere. **Volcanoes** are the landforms created when molten rock escapes from vents in the Earth's surface and then solidifies around these vents. Volcanoes may be active, dormant, or extinct.

Because of its high temperature and relatively low silica content, mafic magma has low viscosity (is highly fluid). It generally erupts (as basaltic lava) relatively quietly, or effusively, because its gases can readily escape and do not build up high pressure. Felsic magma, with its high silica content and relatively low temperature, is highly viscous and generally erupts (as rhyolitic lava) explosively.

The nonexplosive volcanic eruptions characteristic of basaltic lava produce lava flows that, when they solidify, are associated with distinctive features such as pahoehoe- and ʻaʻa-type surface textures, basaltic columns, lava tubes, and pillow structures. The explosive volcanic eruptions characteristic of rhyolitic lavas typically eject **pyroclastic** material—fragments of solidified lava and shattered preex
rock ejected forcefully into the atmosphere. The various
ticles produced when lava cools and solidifies as it falls
to the surface are collectively called **tephra.** Explosive
tion of pyroclastic material is usually accompanied by a
ber of life-threatening effects, such as **pyroclastic flow**
nuée ardentes (high-speed, ground-hugging avalanches
pyroclastic material), and **lahars** (volcanic mudflows).

Nearly all volcanoes have the same two major co
nents: (1) a mountain, or **volcanic cone,** built up of the
ucts of successive eruptions; and (2) a bowl-shaped de
sion, or **volcanic crater,** surrounding the volcano's ve
enough lava erupts to empty a volcano's subterranean
voir of magma, the cone's summit may collapse, form
much larger depression, or **caldera.**

Effusive eruptions, which usually involve basaltic
form gently sloping, broad-based cones called **shield v**
noes. Basaltic magma reaching the surface through lon
ear cracks, or fissures, in the Earth's crust spreads to pr
nearly horizontal lava plateaus.

Explosive **pyroclastic eruptions** involve viscous, us
gas-rich magmas and so tend to produce great amour
solid volcanic fragments rather than fluid lavas. Felsic
olitic) lavas are often so viscous that they cannot flow
a volcano's crater; they therefore cool and harden within
craters to form **volcanic domes.** Ash-flow eruptions occ
the absence of a volcanic cone; they are produced whe
tremely viscous, gas-rich magma rises to just below the
face bedrock, stretching and collapsing it.

The characteristic landform of pyroclastic eruptic
the **composite cone,** or **stratovolcano,** which is compos

alternating layers of pyroclastic deposits and solidified lava. Pyroclastic eruptions may also produce **pyroclastic cones** or **cinder cones,** created almost entirely from the accumulation of loose pyroclastic material around a vent. All pyroclastic-type volcanoes produce steep-sided cones, because the materials they eject—solid fragments and highly viscous lavas—do not flow far from the vent.

Various types of volcanic eruptions are associated with different plate tectonic settings. Explosive pyroclastic eruptions of felsic (rhyolitic) lava generally occur within continental areas characterized by plate rifting or atop intracontinental hot spots. Most intermediate (andesitic) eruptions take place near subducting oceanic plates. Effusive eruptions of (mafic) basalt generally occur at divergent plate margins and above oceanic intraplate hot spots.

Humans can minimize damage from volcanoes by zoning against development in the most hazardous areas, building lava dams, diverting the path of a flowing lava, and learning to predict eruptions accurately. Techniques used to predict eruptions include measuring changes in a volcano's slopes, recording related earthquake activity, and tracking changes in the volcano's external heat flow.

Volcanism is not restricted to the Earth. It has occurred in the past on the Moon, and relatively recent volcanic ac-

End-of-chapter summaries present an overview of the content in narrative form to help students review.

granites are formed within continents by partial melting of the lower portions of the continental crust; these types of igneous rocks are often associated with subduction-produced mountains.

Igneous rocks are also found on the Moon. Lunar igneous rocks differ fundamentally from those on Earth, in that they contain no water and their formation involved neither plate tectonics nor subsurface heat. The Moon's surface consists of highlands composed largely of anorthosite, a coarse-grained plutonic igneous rock, and vast areas of basalt known as maria.

Igneous rocks are valued for the gemstones and precious metals they contain. They are also used for a variety of practical purposes, such as road construction, architectural design, and household abrasives.

Key Terms

igneous rocks (p. 41)
magma (p. 42)
lava (p. 42)
intrusive rocks (p. 43)
plutonic rocks (p. 43)
extrusive rocks (p. 43)
volcanic rocks (p. 43)
peridotite (p. 45)
basalt (p. 45)
gabbro (p. 46)
andesite (p. 46)
diorite (p. 46)
granite (p. 46)
rhyolite (p. 46)
partial melting (p. 46)
Bowen's reaction series (p. 48)
fractional crystallization (p. 49)
plutons (p. 50)
dike (p. 51)
sill (p. 53)
laccolith (p. 53)
lopoliths (p. 53)
batholiths (p. 53)
andesite line (p. 55)

Questions for Review

1. Briefly describe the textural difference between phaneritic and aphanitic rocks. Why do these rocks have different textures?

2. Some igneous rocks contain large visible crystals surrounded by microscopically small crystals. What are these rocks called? How does such a texture form?

3. What elements would you expect to predominate in a mafic igneous rock? In a felsic igneous rock?

4. Name the common *extrusive* igneous rocks in which you would expect to find each of the following mineral types: calcium feldspar; potassium feldspar; muscovite mica; olivine; amphiboles; sodium feldspars. Which *plutonic* igneous rock contains abundant quartz and muscovite mica, but virtually no olivine or pyroxene?

5. What factors, in addition to heat, control the melting of rocks to generate magma?

6. What is the basic difference between the continuous and discontinuous series of Bowen's reaction series?

7. Briefly describe three things that might happen to an early-crystallized mineral surrounded by liquid magma.

8. How do a sill and a dike differ? A laccolith and a lopolith? A lopolith and a batholith?

9. Briefly discuss two specific types of plate tectonic boundaries and the igneous rocks that are associated with them.

10. What is the basic difference between a mid-ocean ridge basalt and an oceanic island basalt?

For Further Thought

1. What type of igneous feature is shown in the photo below?

2. Felsic rocks such as rhyolite often occur together with basaltic rocks near rifting continents. Give one possible explanation for this pairing.

3. Why do we rarely find batholiths made of gabbro?

4. How might the distribution of the Earth's igneous rocks change when the Earth's internal heat is exhausted and plate tectonic movement stops?

5. Why are there virtually no granites or diorites on the Moon? How might small volumes of such felsic rock form under the geological conditions believed to be responsible for the Moon's igneous rocks?

The Key Term list is a tool for quick review and gives the page number for the full discussion, for students who need to reread the material. (The terms also appear in the glossary.)

Questions for Review help students review the factual content of the chapter, and For Further Thought questions encourage them to think critically about the implications of the information they have learned.

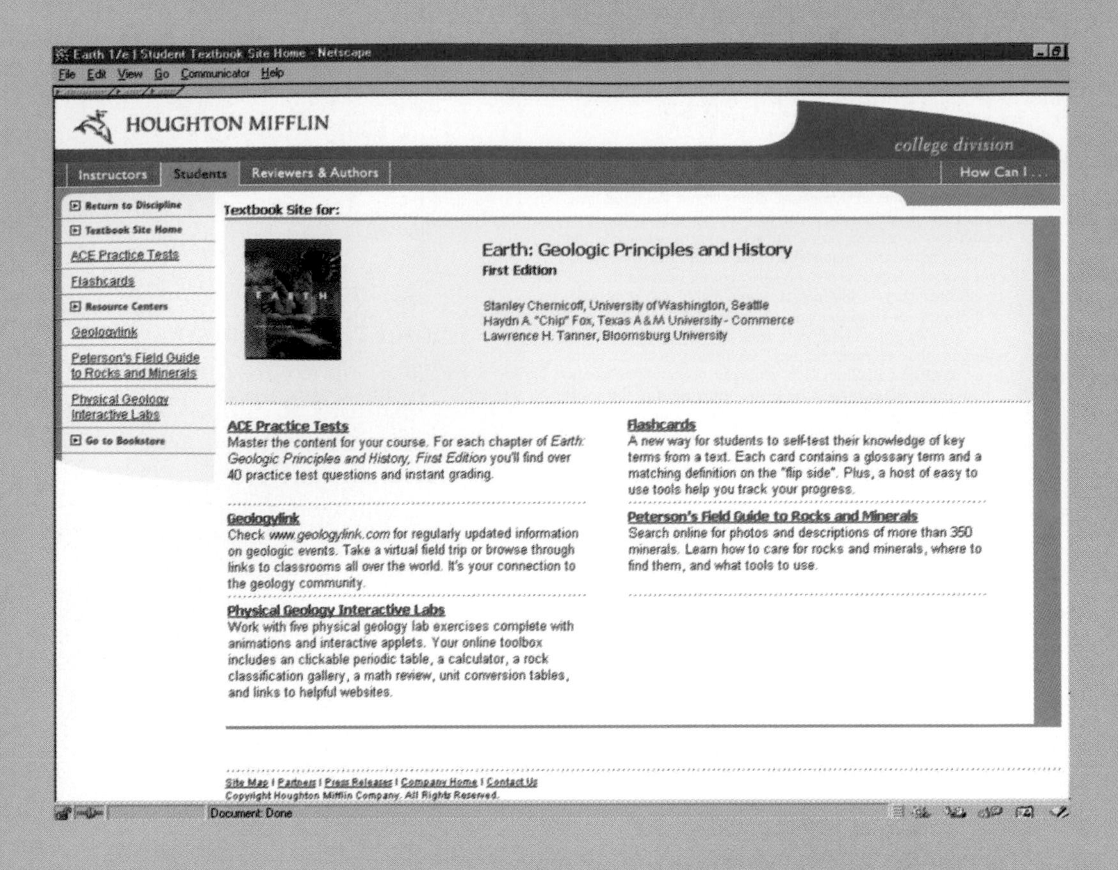

The *Earth: Geologic Principles and History* web site contains a wealth of resources such as ACE practice tests, flashcards of key terms, *Peterson's Field Guide to Rocks and Minerals,* Physical Geology Interactive labs, and a link to www.geologylink.com. Additionally, *Study Guide* material for the last eight chapters of the text can be found on the site. The instructor portion of the site contains downloadable PowerPoint slides, the *Instructor's Resource Manual* for the last eight chapters of the text, and additional course resources.

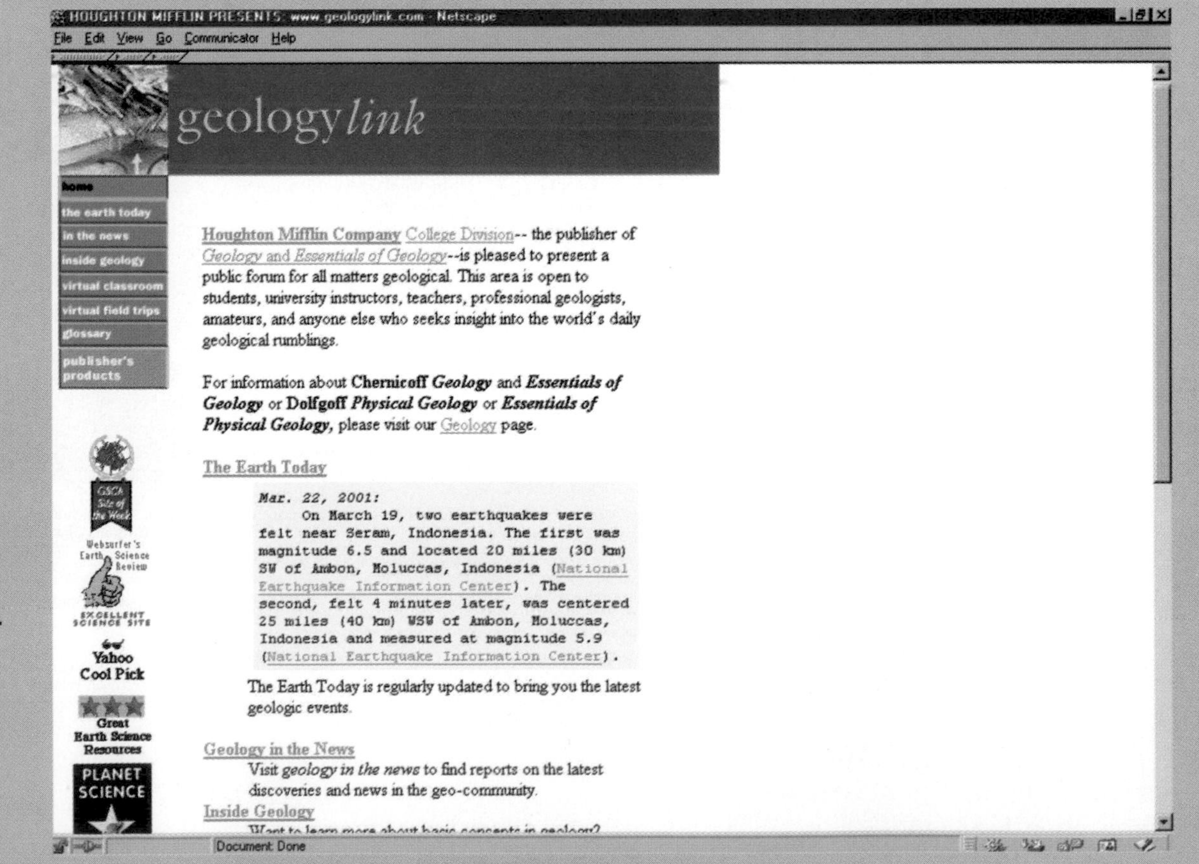

www.geologylink.com is Houghton Mifflin's award-winning site for the geology community. You'll find information about geological events as they happen around the globe, and reports on the latest discoveries and news in the geo-community. Geologylink also features a virtual classroom with links to physical geology courses from around the world, as well as a number of virtual field trips.

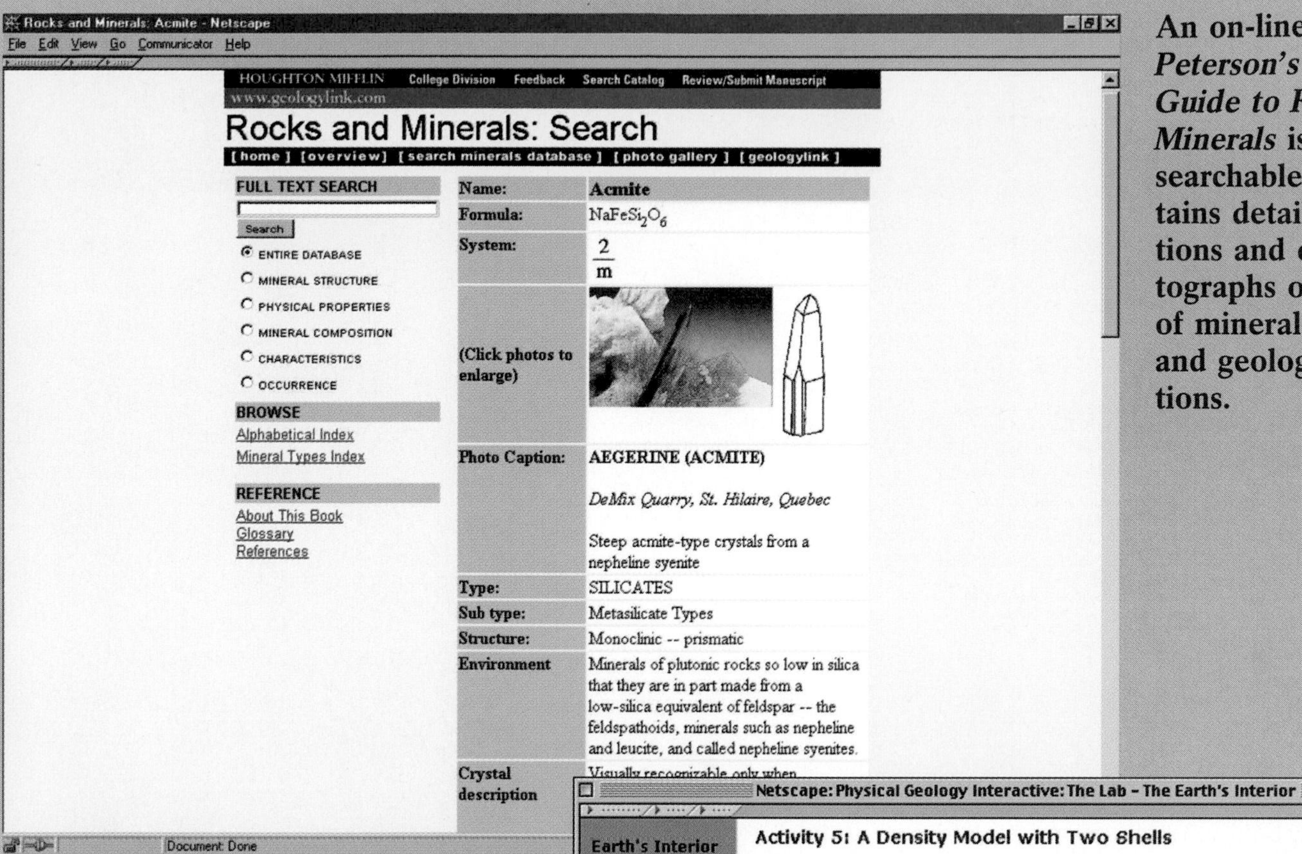

An on-line version of *Peterson's Field Guide to Rocks and Minerals* is fully searchable and contains detailed descriptions and color photographs of hundreds of minerals, rocks, and geologic formations.

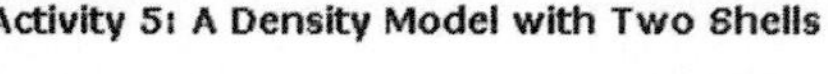

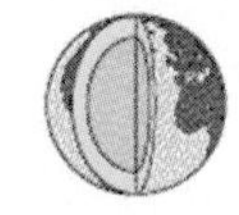

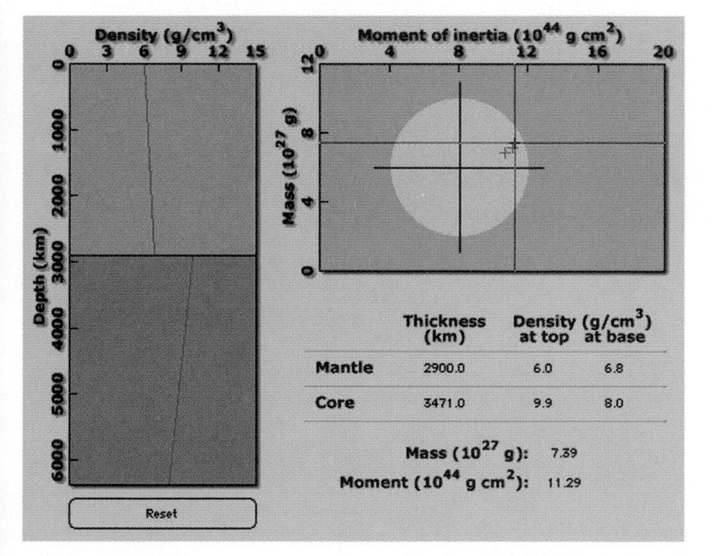

Physical Geology Interactive is an on-line lab manual containing interactive lab exercises, animations, and exercises that can be printed and submitted for grading.

Part 1 Forming the Earth

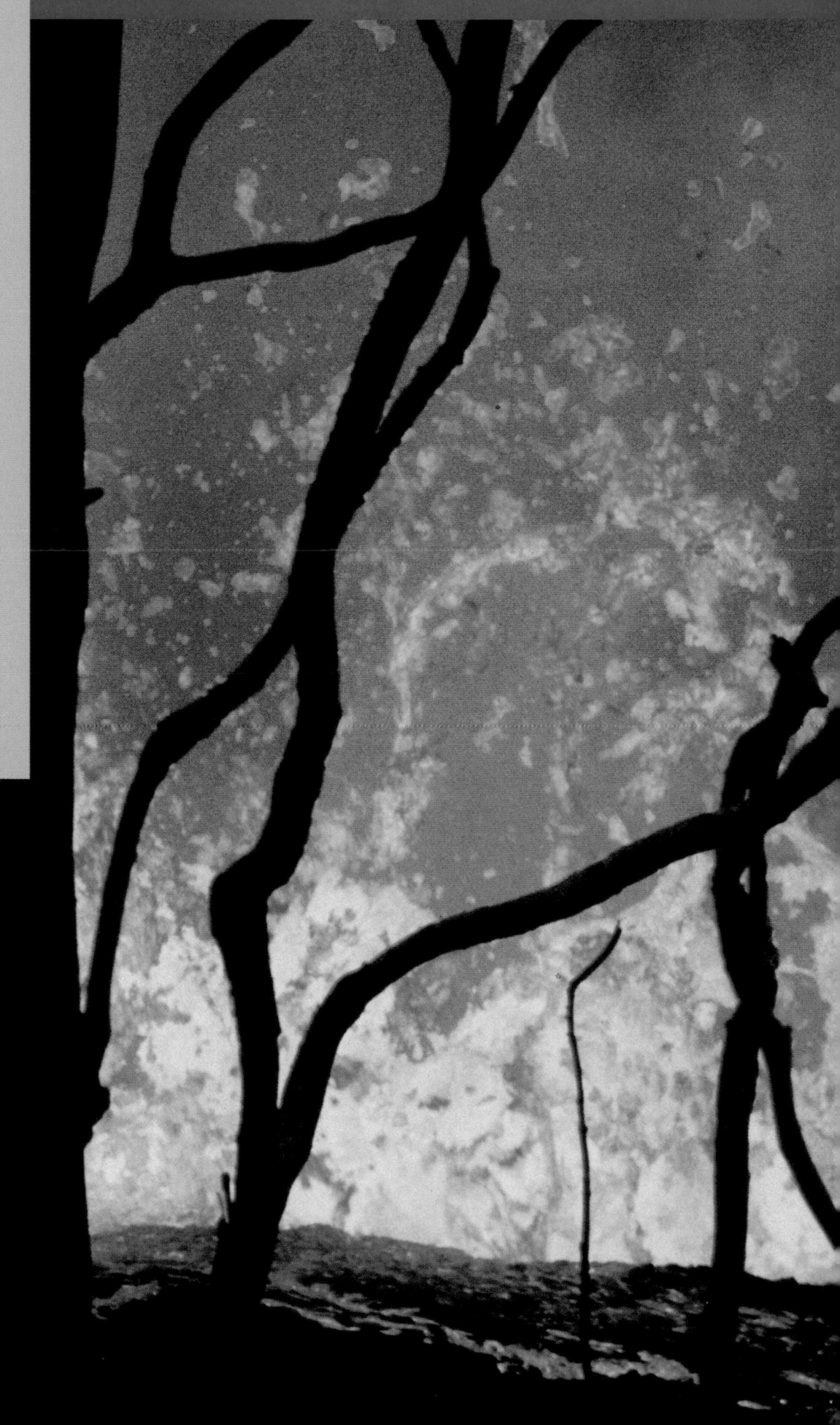

Highlight 1-1
What Caused the Extinction of the Dinosaurs?

The Methods of Science and Geology

The Earth in Space

Rocks and Geologic Time

Plate Tectonics

A Preview of Things to Come

1

A First Look at Planet Earth

At 5:03 P.M. Pacific daylight time on October 17, 1989, baseball fans across North America were settling down in front of their television sets to watch Game Three of the World Series from San Francisco. Minutes later, violent movement along a small segment of California's San Andreas fault (Fig. 1-1) had caused widespread destruction, taking the lives of scores of Bay Area residents and injuring hundreds more. Instead of baseball, millions viewed live broadcasts of grim scenes, including collapsed buildings and freeways (Fig. 1-2). Four years later, at 4:31 A.M. Pacific standard time on January 17, 1994, Southern Californians were jolted awake by a powerful earthquake that took 57 lives, buckled numerous freeways, and proved to be one of the most expensive natural disasters ever in the United States, with estimated cleanup and repair costs of more than $15 billion. Exactly one year later, the effects of the 1989 Loma Prieta and 1994 Northridge earthquakes in California were put into a new perspective by another earthquake—this one more than 9000 kilometers (5700

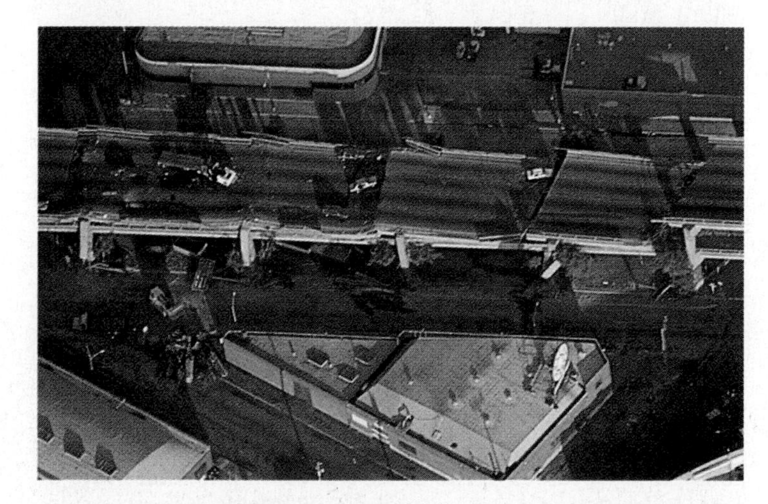

Figure 1-2 The collapse of the Nimitz Freeway in Oakland, California, during a major earthquake along the San Andreas fault on October 17, 1989. The San Andreas fault, a fracture in the Earth's crust that cuts northwest-southeast across much of California, is responsible for some of North America's most powerful earthquakes.

Figure 1-1 The San Andreas fault, as seen from the air over Carrizo Plain in California.

Highlight 1-1 What Caused the Extinction of the Dinosaurs?

Paleontologists (geologists who study ancient life forms) have long wondered what might have caused more than 75% of all the forms of life then on Earth to vanish about 65 million years ago. The most dramatic loss involved the extinction of the dinosaurs, a group of animals that had roamed the planet for 150 million years, but numerous other life forms vanished as well—large and small, water- and land-dwelling, plant and animal. Many species—whether living in freshwater lakes, in rivers, in saltwater oceans, or on land—became extinct at roughly the same time.

Some early hypotheses focused on only one kind of organism to explain these extinctions. Some proposed that epidemic diseases eliminated dinosaur populations or that the rise of egg-stealing mammals ravaged dinosaur nests. But neither of these hypotheses accounted for the loss of two-thirds of all marine animal species, which led some scientists to propose that the oceans became lethally salty (though this idea did not explain why some marine creatures survived). To explain the extinction of gigantic terrestrial reptiles, tiny marine organisms, and many life forms in between, a number of hypotheses invoked global environmental change. Did the Earth suffer from a period of drastic cooling 65 million years ago? Did a shift in the planet's protective magnetic field allow harmful solar radiation to reach land and sea, eliminating a wide variety of life forms? Did a nearby star explode, bathing the Earth in cosmic radiation? Surely, each of these events would have affected all life on Earth simultaneously. Why, then, did 25% of the planet's species remain unaffected?

Several hypotheses agree that wholesale extinction followed some catastrophic disruption of the global food chain. One group of scientists has proposed that massive volcanic eruptions of India's Deccan plateau may have been such an event. The basalts from these eruptions have been dated to 65 million years ago, coinciding perfectly with the extinction of the dinosaurs. The eruptions consisted of hundreds of lava flows covering an area of 10,000 square kilometers (3861 square miles) that produced more than 10,000 cubic kilometers (2390 cubic miles) of basalt. It is postulated that the eruptions sent a cloud of volcanic ash and gas around the Earth, blocking out sunlight, cooling the planet, and leading to a worldwide decline in vegetation, including microscopic marine plants. Without the plants on which their diets were based, many plant-eating animals would have died, and their extinction would in turn have wiped out the meat-eaters, such as *Tyrannosaurus rex,* that were their predators.

Another group of scientists, led by geologist Walter Alvarez and his father, Nobel prize–winning physicist Luis Alvarez, has proposed another scenario: A meteorite at least 10 kilometers (6 miles) in diameter plowed into the Earth, releasing a shower of pulverized rock into the atmosphere. The resulting dust veil would have blocked out sunlight (in much the same way as volcanic ash would have), cooled the planet, and led to an "impact winter" that may have lasted for decades—long enough to devastate the global food chain. The strongest evidence to support this impact hypothesis is a 2.5-centimeter (1-inch)-thick

miles) away near Kobe, Japan—that killed more than 5000 people, injured 30,000, and left 300,000 homeless.

Why do areas such as California and Japan suffer from periodic and often devastating earthquakes while other areas are spared? The answer to this question, as well as to questions about why volcanoes, landslides, and other catastrophic events occur, can be found in the science of geology. **Geology** is the scientific study of the Earth: the materials that compose it; the processes, such as mountain building and the creation of ocean basins, that shape its surface; and the events, such as floods and glaciation, that sculpt its landscapes. Geologists examine the origin of the Earth and its evolution through its 4.6-billion-year history, and even the geological processes of the other planets in our solar system.

Everything we use comes from the Earth, so geology has an enormous practical impact on our daily lives. Through geological knowledge we are able to locate natural resources, such as the oil, gas, and coal that fuel our cars and heat our homes, the iron and other metals upon which so much of our civilization relies, and even new sources of clean groundwater that are essential to life and agriculture in many areas. Geological study also helps us to predict and avoid some of nature's life-threatening hazards—for example, by identifying slopes that are too unstable to support buildings, warning us away from eroding shorelines, or even predicting where and when earthquakes might occur. In addition, geologists probe the most fundamental mysteries of our planet: How old is the Earth? How did it form? When did life first appear? Why do some areas suffer from devastating earthquakes or volcanoes, while others are spared? Why are some regions endowed with breathtaking mountains and others with fertile plains?

We begin our study of geology by describing how the science of geology operates and by introducing some basic concepts and standards that underpin this discipline. We discuss how our planet may have formed and speculate about how it has changed over the millennia. Finally, we examine some of the Earth's large-scale geological processes and determine how geologists deduced the nature of these processes.

layer of clay found around the world in rocks that date from approximately 65 million years ago (Fig. 1). The clay contains iridium, an element that is extremely rare in rocks of terrestrial origin, but quite common in meteorites. Mineral grains shattered by very high pressures—as would occur if they had been struck by a meteorite—have also been found at the proposed impact sites. The Alvarezes and their associates contend that the iridium-rich layer resulted from the global fallout of pulverized meteorite dust. Fossils of numerous species, including many now-extinct organisms, have been found in the rocks that formed just before the iridium-rich layer was deposited, whereas only about one-fourth as many species are represented in the rocks that formed just after this layer was deposited. This evidence suggests that many extinctions occurred during the time of deposition.

Just as hypotheses may be discarded, modified, or elevated to theory status, they are also sometimes combined. One group of scientists has recently proposed that the Earth was indeed struck by a meteorite 65 million years ago somewhere in the Western Hemisphere, and that reverberations from the impact initiated massive volcanism on the opposite side of the globe. The material spewed into the atmosphere by both events may have combined to devastate the Earth's food chain, thus bringing about the demise of the dinosaurs.

As yet, no extinction hypothesis has achieved theory status. Analysis of the Earth's 65-million-year-old deposits continues today, as scientists seek to document further the proportions of organisms that became extinct at that time and search for additional evidence of a meteorite strike or of a catastrophic volcanic eruption that coincides with the time of the extinctions.

Figure 1 An iridium-containing layer of clay (marked by coin) found by Walter Alvarez in Gubbio, Italy. The Alvarezes believe that this clay, which is found around the world in rock of this age, may have been deposited after a meteor impact about 65 million years ago.

The Methods of Science and Geology

Scientists make one basic assumption: The world works in an orderly fashion in which natural phenomena will recur given the same set of conditions. As scientists understand it, every effect has a cause.

The Scientific Method

The principal objective of science is to discover the fundamental patterns of the natural world. In trying to find the reasons underlying natural phenomena, scientists use a distinctive strategy called the **scientific method.** First, they gather all available information bearing on their subject, such as measurements and descriptions taken in the field and the results of laboratory experiments. Then, they develop a **hypothesis,** a tentative explanation that fits all the data collected and is expected to account for future observations as well. Often, a number of different competing hypotheses are proposed to explain the same set of data. Highlight 1-1, for example, describes the various hypotheses put forth to explain the mysterious extinction of the dinosaurs and many other species 65 million years ago.

Hypotheses are tested over time as scientists conduct further experiments and make further observations. If a hypothesis does not explain subsequent findings, it must be modified or abandoned. The history of science is littered with disproved hypotheses that were once quite popular—such as the suggestion that the Earth is the center of our solar system. A hypothesis that is repeatedly confirmed by extensive observation and experimentation is retained and may become a **theory,** an explanation that has remained consistent with all the data and gained wide acceptance within the scientific community.

Even after a hypothesis survives testing and becomes a theory, sometimes new data become available—perhaps as a result of updated technology—that are not consistent with the theory. Scientists then propose new hypotheses, modifying or completely replacing the established theory. A theory that continues to meet rigorous testing over a long period of time may be declared a **scientific law.** For example, it has been observed repeatedly that when an object is dropped, it falls toward the Earth's surface. The invariability of this observation has led scientists to accept the *law of gravity* as a scientific law.

Geologists, like other scientists, use field observations and laboratory experimentation in much of their work. However, because most geological processes are imperceptibly slow and their scale unimaginably large from a human perspective, they can't always test hypotheses through direct observation or experimentation in the same way that chemists or physicists can. To supplement their field and laboratory work, geologists sometimes use scaled-down models to study large-scale geological phenomena or rely on the power of computers to create mathematical models.

The Development of Geological Concepts

Almost two centuries of observation and hypothesis formation and testing have contributed to our current understanding of how the Earth developed. Prior to the mid-eighteenth century, the common belief was that the Earth's geological evolution had taken place through a series of immense worldwide upheavals such as volcanic eruptions, monumental earthquakes, and worldwide floods. This belief, called **catastrophism,** was called upon to explain the existence of mountains, valleys, fossils, and all other geological features found on the Earth. Inherent in catastrophism was the belief that the Earth was only a few thousand years old, a concept intrinsic to many Christian theologies.

During the latter part of the eighteenth century, the Scottish naturalist James Hutton (1726–1797) proposed that the processes that anyone could see changing the Earth in small ways during his or her lifetime must have operated in a similar manner throughout the planet's history. His hypothesis, called **uniformitarianism,** proposed that current geological processes could be used to explain long-past geological events. Hutton recognized that slow processes, such as rivers cutting through valley floors and loose soil creeping down gentle slopes, acting over a vast amount of time, may have had a greater effect on the Earth than did occasional catastrophic events (Fig. 1-3). Because it assumed that the Earth was much older than the few thousand years most people believed it to be, Hutton's hypothesis met with great resistance. By the 1830s, however, after much debate, uniformitarianism had prevailed over catastrophism. Its acceptance has been hailed as the birth of modern geology, and James Hutton is widely considered to be the "father of modern geology."

James Hutton maintained that "the present is the key to the past." Indeed, geologists today recognize that the Earth's present appearance results from millions of years of the same physical processes, although probably acting at varying rates.

(a)

(b)

Figure 1-3 Gradual change of the Earth. Even processes that occur at very slow rates can change the Earth's appearance dramatically over long periods of time. These two photos of the Grand Canyon, taken from the same perspective 100 years apart, show little geological difference between the earlier scene **(a)**, photographed in 1873, and the later scene **(b)**, photographed in 1972. Fossils of ancient marine creatures found in some rocks of the Grand Canyon, however, show that these rocks—found today near the top of a hot, dry plateau 2300 meters (7500 feet) above sea level—once lay at the bottom of an ocean. Over millions of years, the mud at the bottom of this ocean gradually solidified into rock, was uplifted to its present position, and was cut through and exposed by the Colorado River.

They also know, however, that some geological events are indeed catastrophic, and that much geological change does occur during brief spectacular events. A great earthquake may shift a land area more than 6 meters (20 feet) in a single moment. In 1989, Hurricane Hugo eroded more of the Carolina coast in one day than had the preceding century of slow, steady wave action. Both slow, consistent processes and catastrophic events continuously shape our planet.

Modern geology is concerned with more than just the surficial features of the Earth. It is also concerned with the formation, subsequent changes, numerous characteristics, and fantastic variety of rocks and minerals. It is concerned with the deep unseen interior of the Earth. It is also concerned with the Earth's characteristics relative to other planets in our solar system and with the very origin of the universe.

The Earth in Space

The Earth is a slightly flattened sphere with an average radius of 6371 kilometers (3957 miles), orbiting approximately 150 million kilometers (93 million miles) from the medium-sized star we call the Sun. Our Sun is only one of about 100 billion stars in the Milky Way galaxy, a pancake-shaped cluster of stars that itself is only one of about 100 billion such galaxies in the observable universe. Despite Earth's relative insignificance compared to the universe as a whole, it is perfectly positioned to receive just the right amount of the Sun's radiant energy to support life. Because of its composition and geologic past, the Earth has manufactured a watery envelope and protective atmosphere on which countless living species have relied for millions of years. But how did the Earth become what *may* be the only life-sustaining planet in the solar system?

The Probable Origin of the Sun and Its Planets

Cosmologists (scientists who study the origin of the universe) have proposed that the universe began as a very small, very hot volume of space containing an enormous amount of energy. Many scientists believe that the birth of all the matter in the universe occurred when this space expanded rapidly with a "Big Bang" roughly 12 billion years ago. Immediately after the Big Bang, they suggest, the universe began to expand and cool, which it continues to do today. About a million years after the Big Bang, when the universe had cooled sufficiently to allow the first atoms to form, the universe consisted of about 75% hydrogen gas and 25% helium gas, just as it does today. As the universe continued to expand, pockets of relatively high gas concentrations began to form because of gravitational attraction among the gas particles. Where enough gas gathered, the resulting gas clouds collapsed inward from the force of gravity and created galaxies and clusters of galaxies.

Within each galaxy, such as our own Milky Way, some gas clouds collapsed further to form stars. Even today, stars continue to be born in this way in all galaxies, including our own. (Through a telescope, you can see a "star nursery" in the belt of the constellation Orion, for example.) As the gas within each star collapsed under gravity, sufficient heat was generated to fuse together particles within the core, a phenomenon known as *nuclear fusion.* Such nuclear fusion produces the light we see when we look at stars, including the Sun.

Stars are not only born, but also die—some slowly and some rapidly. Stars die when they begin to exhaust their supply of nuclear fuel and collapse under their own gravitational force. A star that is dying very rapidly is called a *nova* (Latin for "new") because it appears as a very bright new star in the heavens. Dying stars are important because they generate so much heat that new nuclear reactions occur, producing the nuclei of the heavier elements of which the Earth and other planetary bodies are composed.

Long after our galaxy was created, the remnants of earlier stars that had died contributed to the gas cloud, or *nebula,* that eventually developed into our solar system (Fig.1-4). Our nebula was probably originally dispersed across a vast area of space, extending well beyond what would become the orbit of our solar system's outermost planet, Pluto. About 5 billion years ago, this nebula began to collapse inward, perhaps due to a shock wave from a nearby exploding star. As its component materials were drawn by gravity toward its center, they collided in nuclear reactions and generated heat in the same way as other stars, forming the infant Sun.

As heat became concentrated in the center of this new star, material in the outer nebula surrounding it began to cool and condense into infinitesimally small grains of matter. Uncondensed substances nearby were swept outward by strong solar winds, consisting of streams of matter and energy that flowed from the infant Sun. In this way, the first solid materials to form in our solar system became separated into a hot inner zone of denser substances, such as iron and nickel, and a cold outer zone of low-density gases, such as hydrogen and helium. Ultimately, this compositional partitioning would evolve into the four rocky inner planets and the five gaseous outer planets.

As the first bits of matter condensed, they collided and coalesced, forming aggregates that grew to a few kilometers or larger in diameter. These planetary seeds, or *planetesimals,* formed the cores of the developing planets. Cosmologists originally believed that this process of planetary growth, or *accretion,* was slow and gradual, much as one might create a large aluminum-foil ball by the steady addition of small lumps. Recently, however, our view of planetary accretion has changed drastically. Cosmologists now believe that as they grew, huge planetesimals—easily the size of Mercury or even Mars—collided violently. Such violent collisions ejected great masses of molten material into space, perhaps forming some of the moons that today orbit the planets of our solar system.

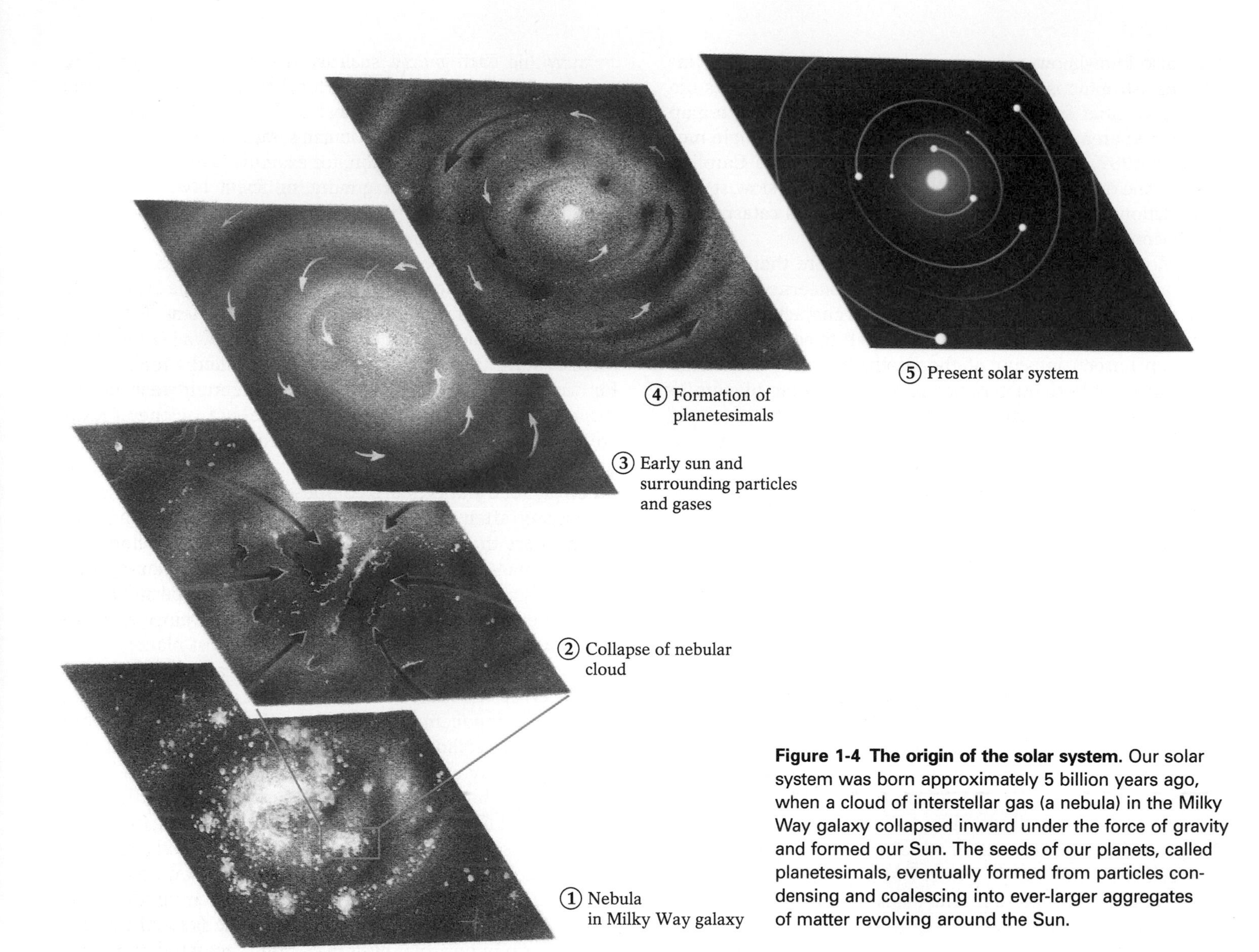

Figure 1-4 The origin of the solar system. Our solar system was born approximately 5 billion years ago, when a cloud of interstellar gas (a nebula) in the Milky Way galaxy collapsed inward under the force of gravity and formed our Sun. The seeds of our planets, called planetesimals, eventually formed from particles condensing and coalescing into ever-larger aggregates of matter revolving around the Sun.

Intense solar radiation warmed the four planets closest to the Sun, causing their surface temperatures to rise. Nearly all of their lighter gases vaporized and were carried away by solar winds. The matter remaining in these four small, dense, inner planets—Mercury, Venus, Earth, and Mars—consisted primarily of iron, nickel, and silicate minerals, which contain a large amount of silicon and oxygen. Much farther from the Sun's heat, the outer planets—Jupiter, Saturn, Uranus, Neptune, and Pluto—formed primarily from the now-frozen lighter gases, hydrogen, helium, ammonia, and methane.

The Earth's Earliest History

The Earth, the largest of the four inner planets, began as a mostly solid, homogeneous body of rock and metal. This temporary state was later changed by the extreme violence and chaos that characterized the Earth's first 20 million years. Proto-Earth's collisions with other planetesimals converted the enormous energy of motion to thermal energy upon impact. Some of this energy was retained and "buried" by succeeding collisions and accretions. "Compressional heating" also resulted from the accumulation of the mass of overlying rocks. In addition, *radiogenic* heating occurred as the atoms of radioactive substances, such as uranium, released heat as their nuclei split apart, a process known as *fission.* The heat from these two sources caused the planet's internal temperature to rise tremendously and set in motion the process that created a layered Earth.

During the Earth's first 10 to 20 million years, the planet's internal temperature rose to the melting point of iron. As a result, much of the iron liquefied. Because it was more dense than the surrounding materials, the iron sank to the proto-Earth's center by the pull of gravity. As it sank, less dense materials rose and became concentrated closer to the planet's surface. Thus the matter that had originally made up a homogeneous Earth became separated into three major concentric zones of differing densities. (This separation process is called **differentiation.**) The densest materials, probably iron and nickel, formed a core at the planet's center. Lighter materials, composed largely of silicon and oxygen as well as other relatively light elements, formed the Earth's outer layers (the mantle and crust). The arrangement of these three layers is somewhat like that of a hard-boiled egg with its thin shell, extensive white, and central yolk. The egg model, however, does not show a number of important sublayers that are fundamental to our understanding of our dynamic Earth. Even lighter materials—gases that had been trapped in the interior—escaped, combining to form the Earth's first atmosphere and oceans.

A Glimpse of the Earth's Interior As the Earth differentiated into its major concentric layers (Fig. 1-5), some upwelling material reached the surface, where it cooled and solidified, forming the Earth's earliest **crust.** Among these low-density substances were oxygen and silicon, which combined to form the silicate minerals that abound in the Earth's crust and upper mantle. Some heat-producing radioactive substances, such as uranium and thorium, also moved toward the surface; because of the heat radiating from these elements, crustal rocks are repeatedly remelted and re-formed into a wide variety of rock types.

Underlying the crust is the **mantle,** a thick layer of denser rocks. The outer 100 kilometers (60 miles) of the Earth, encompassing both the crust and the uppermost portion of the mantle, is a solid, brittle layer known as the **lithosphere** ("rock layer," from the Greek *lithos,* meaning "rock"). Underlying the lithosphere is the **asthenosphere** ("weak layer," from the Greek *aesthenos,* meaning "weak"), a zone of heat-softened rock located in the upper mantle roughly 100 to 350 kilometers (60–220 miles) beneath the Earth's surface. Although it remains solid, the heat-softened rock of the asthenosphere actually flows slowly—a phenomenon that drives much of the planet's geological activity. The lithosphere and asthenosphere are where such large-scale geological processes as mountain building, volcanism, earthquake activity, and the creation of ocean basins originate.

Below the mantle and at the Earth's center is the **core,** the densest layer of all. The core is divided into a liquid outer core and a solid inner core, both consisting primarily of iron and nickel.

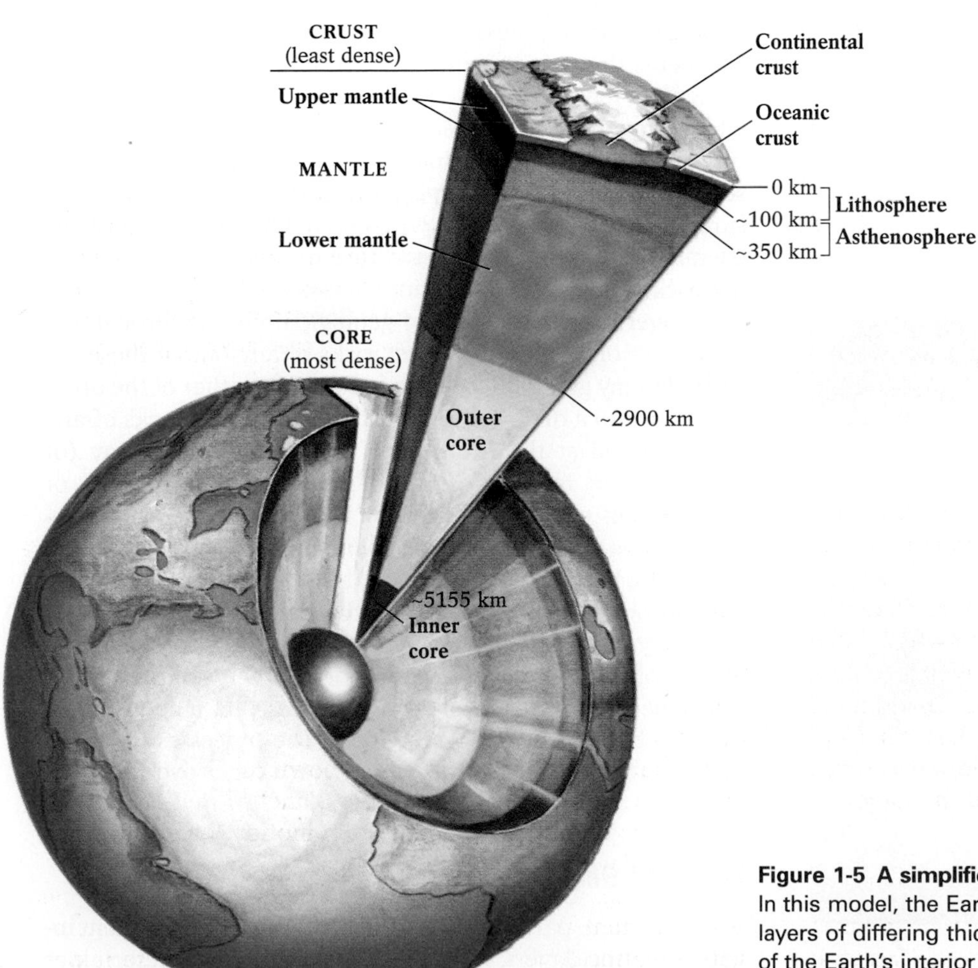

Figure 1-5 A simplified model of the Earth's interior. In this model, the Earth is composed of concentric layers of differing thicknesses and densities. A slice of the Earth's interior reveals a thin crust, a massive two-part mantle, and a two-part core.

The Origin of the Moon The birth of our Moon has sparked lively debate for centuries. Did it form as a companion planet coalescing independently from the solar nebula at the same time as Earth? Did it form elsewhere, only to be drawn into Earth's orbit by our planet's relatively strong gravity? Or was the Moon once part of the Earth?

The answer may lie in the Moon's composition. It is 36% less dense than the Earth and apparently contains much less iron. This difference rules out independent accretion from the solar nebula, for if the Moon did form in the same way as the

Earth, its composition would be similar. The Moon's composition, confirmed in part by the rock-collecting efforts of U.S. Apollo astronauts, is actually quite similar to that of the Earth's mantle, a fact that has led many scientists to suggest that the Moon was formed in a cataclysmic collision between the Earth and another planetesimal.

By roughly 4.55 billion years ago, the Earth had probably attained much of its current size and had become layered, with most of its iron having migrated toward the center to form the core. With the Earth's relatively large gravitational pull, it may have attracted a Mars-sized planetesimal. With the planetesimal traveling toward the Earth perhaps as fast as 14 kilometers per second (31,500 miles per hour), its impact would be quite literally Earth shattering (Fig. 1-6). At the moment of impact, the Earth's young atmosphere would have been blown away, replaced by a rain of molten iron blobs, remnants of the planetesimal's iron core. Such a collision would have vaporized much of the crust and mantle of both the Earth and the planetesimal. Jets of the vaporized crust and mantle would be shot into orbit around the Earth, where the material could eventually coalesce to form the Moon.

Where are the "wounds" of this great collision? Unfortunately, the Earth's dynamic internal processes—related to volcanism, earthquakes, and mountain building—would have eradicated much of the evidence, and erosion would have eliminated the rest. A search for the evidence of the greatest collision in the Earth's history is unlikely to yield a clue, and the story of the formation of the Earth's Moon must remain only a hypothesis.

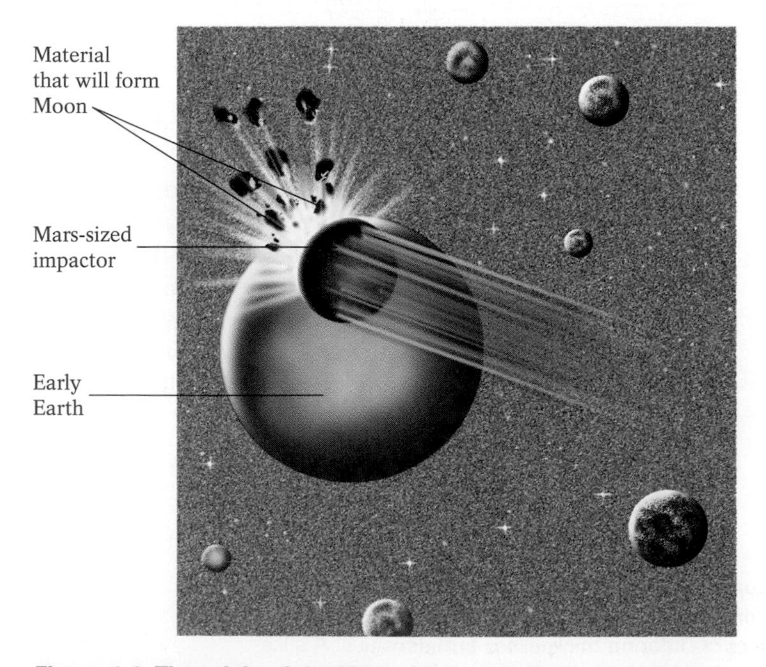

Figure 1-6 The origin of the Moon. Many scientists now believe that a catastrophic impact between the proto-Earth and a Mars-sized planetesimal spawned the Earth's Moon.

Rocks and Geologic Time

The phrase *geologic time* refers to the time elapsed between the formation of the Earth and the beginning of human history—almost all of the planet's 4.6 billion years of existence. This phrase can also refer to the long spans of time over which geological processes such as mountain building occur. In human terms, the extent of geologic time is almost inconceivably long, and the processes that occur over geologic time are almost inconceivably slow. Fortunately, rocks can bear witness to the effects of even the slowest processes and the oldest events, and almost all of our knowledge about our planet's past reflects the fact that it was preserved, in some way, in rock.

Rock Types and the Rock Cycle

A **rock** is a naturally formed aggregate of one or more minerals. Three types of rocks exist in the Earth's lithosphere and at its surface, with each type reflecting a different process of origin. **Igneous rocks** form from the cooling and crystallization of molten material that has migrated from the Earth's interior to, or just beneath, the Earth's surface. **Sedimentary rocks** form when preexisting rocks become broken down into fragments that accumulate and then become compacted and cemented together. They may also form from the accumulated and compressed remains of certain plants and animals, or from chemical precipitates of materials previously dissolved in water. **Metamorphic rocks** form when heat, pressure, or chemical reactions with circulating fluids change the chemical composition and structure of any type of preexisting rock in the Earth's interior.

Over the great extent of geologic time and through the dynamism of Earth's processes, rocks of any one of these basic types may gradually be transformed into either of the other types, or into a different form of the same type. Rocks of any type exposed at the Earth's surface can be worn away (or *weathered*) by rain, wind, crashing waves, flowing glaciers, or other means, and the resulting fragments transported elsewhere to be deposited as new *sediment;* this sediment might eventually become new sedimentary rock. Sedimentary rocks may become buried so deeply in the Earth's hot interior that they may be changed into metamorphic rocks, or they may melt into magma and eventually cool and recrystallize to form new igneous rocks. Under heat and pressure, igneous rocks can also become metamorphic rocks. The processes by which rocks can form and re-form into different types over time are illustrated in the **rock cycle** (Fig. 1-7).

Time and Geology

An important part of reconstructing our geologic past involves dating layers of rock. *Relative dating* determines which rocks are older than others by referring to spatial relationships between the rocks. For example, the geologic

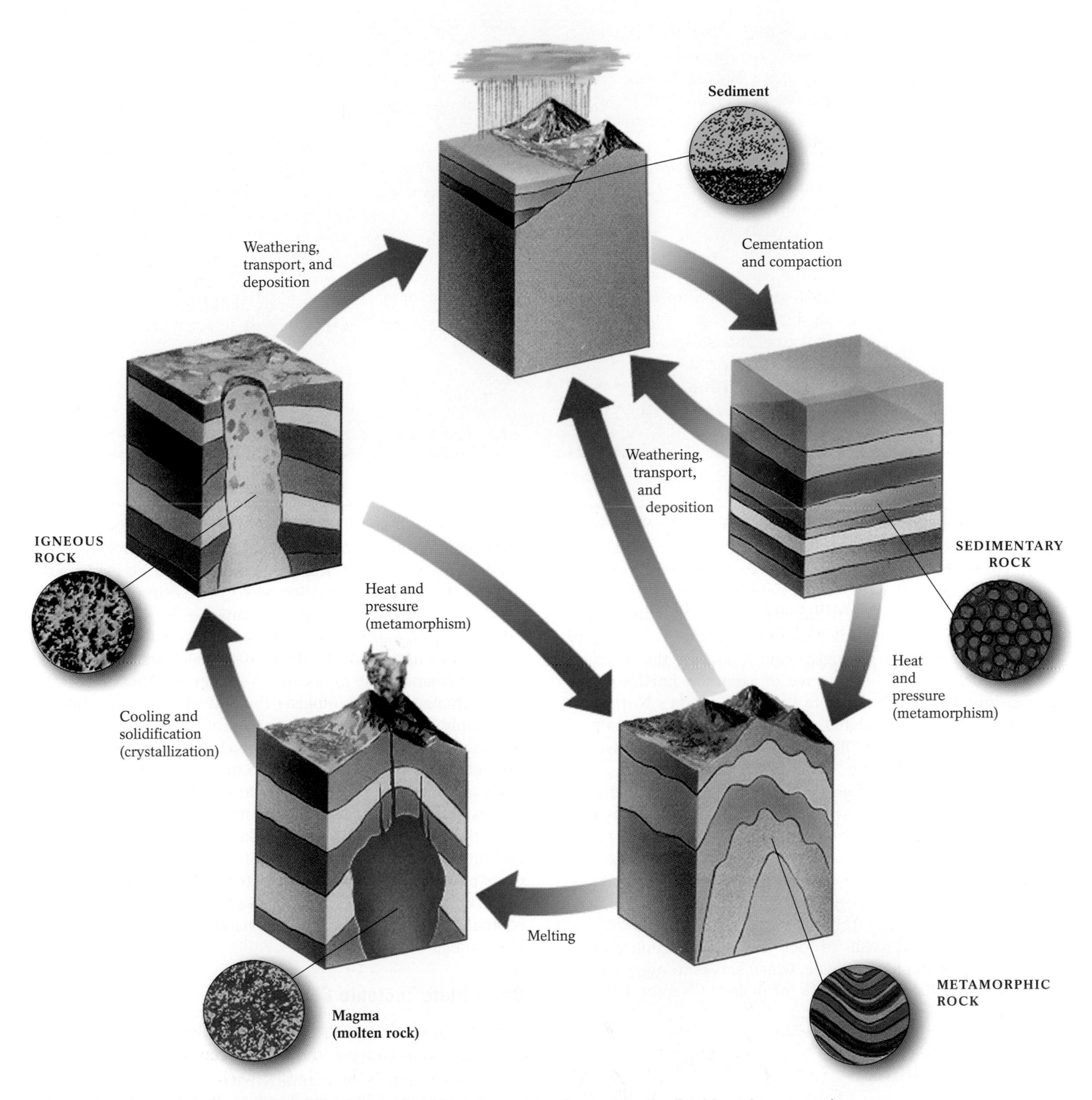

Figure 1-7 The rock cycle. This simplified scheme illustrates the variety of ways that the Earth's rocks may evolve into other types of rocks. For example, an igneous rock may weather away and its particles eventually consolidate to become a sedimentary rock. The same igneous rock may remain buried deep beneath the Earth's surface, where heat and pressure might convert it into a metamorphic rock. The same igneous rock, if it is buried even deeper, may actually melt to become magma—which may eventually recool and solidify to form a new igneous rock. There is no prescribed sequence to the rock cycle. A given rock's evolution may be altered at any time by a change in the geological conditions around it.

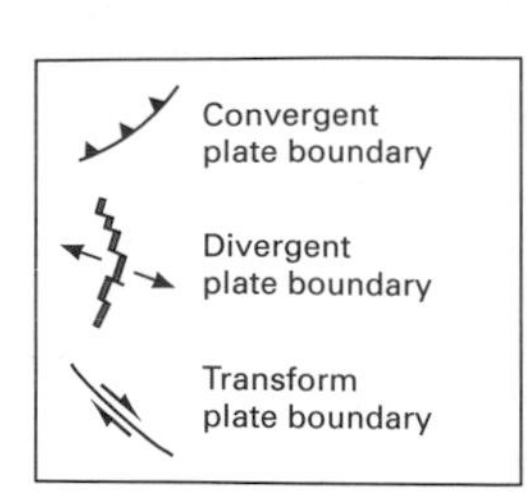

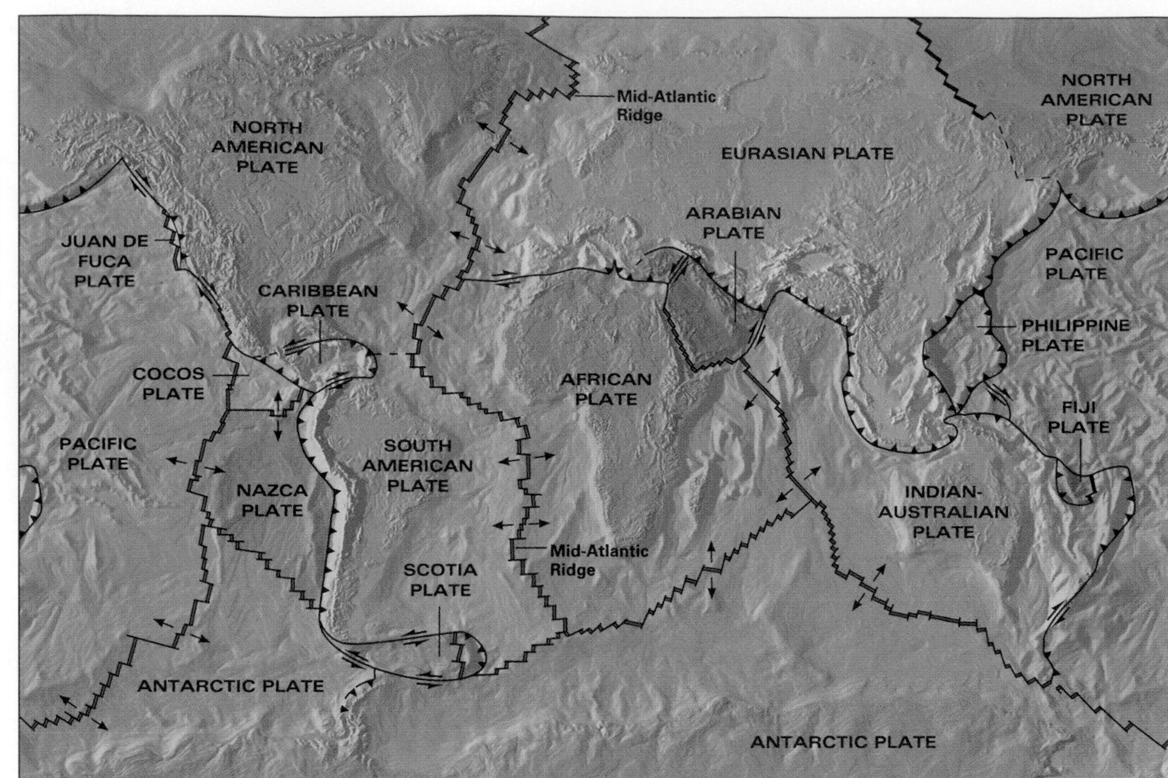

Figure 1-8 A world map showing the Earth's tectonic plates. Note that some plates, such as the North American plate, are composed of both continental and oceanic lithosphere. The Pacific plate is made up almost exclusively of oceanic lithosphere.

principle of superposition states that, where layers of sedimentary rocks have not been disturbed since their deposition, younger rocks overlie older rocks. Equally important is *absolute dating* of rocks, which determines a rock's age in years. Using rock-dating technology developed during the twentieth century, which is based on the constant decay of radioactive elements, geologists can determine the *absolute* ages of rocks with a high degree of accuracy. Earth's oldest rocks, found near Yellow Knife Lake in Canada's Northwest Territories, have been dated using the known decay rate of uranium. They are 3.96 billion years old! We will discuss the various dating methods in detail in Chapter 8.

Our knowledge of past events also relies a great deal on the presence of *fossils*—traces or remains of long-dead plants and animals—in some rocks. These can tell us not only about the structures and activities of extinct organisms, but also—depending on their relative placement in the rock—approximately when the organisms lived. In conjunction with the various rock-dating techniques, the information provided by the sequence of fossils in rock has allowed geologists to develop the geologic time scale, which serves to organize the history of the Earth and its life forms (see Chapter 19).

Plate Tectonics

Today's geologists understand that the Earth behaves in many ways as a single, dynamic system. In earlier centuries, however, geologists were unable to combine the explanations for various geological phenomena, such as the existence of mountain ranges, ocean basins, earthquakes, and volcanoes, into a general explanation of how the Earth functions. Their hypotheses were tailored to specific locations and could not be applied elsewhere. In short, geologists did not have an underlying theory by which *all* geological phenomena could be explained.

In the 1960s, an exciting new hypothesis emerged that provided an elegant unifying explanation for all geological processes, past and present. This hypothesis, called **plate tectonics,** revolutionized the way that geologists viewed the world as dramatically as, a century earlier, the theory of evolution changed how biologists thought about living things. After only a few decades of observation and testing, the hypothesis of plate tectonics has become a widely accepted theory because it provides answers to questions that earlier hypotheses could not resolve. For example, it has enabled us to understand processes such as mountain building, predict such potential catastrophes as earthquakes and volcanic eruptions, and find underground reservoirs of oil, natural gas, and precious metals. Finally, the plate tectonic theory enables us to fit our observations about the ancient past into the same conceptual framework as our understanding of the geological phenomena occurring today.

Basic Plate Tectonic Concepts

The theory of plate tectonics relies on four basic concepts:

1. The outer portion of the Earth—its crust and uppermost segment of mantle (that is, its lithosphere)—is composed of large rigid units called plates.
2. The plates move slowly in response to the flow of the heat-softened asthenosphere beneath them.
3. Most of the world's large-scale geological activity, such as earthquakes and volcanic eruptions, occurs at or near plate boundaries.

CONTINENTAL CRUST

Stressed continental plate begins to rift

Rising currents in asthenosphere

(a)

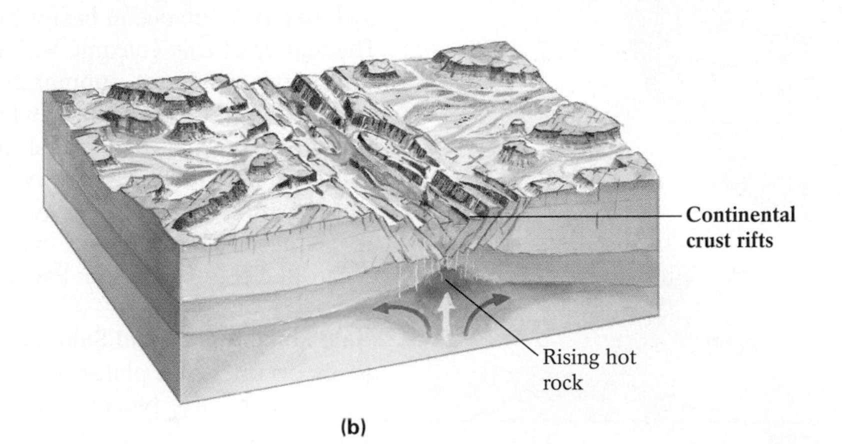

(b)

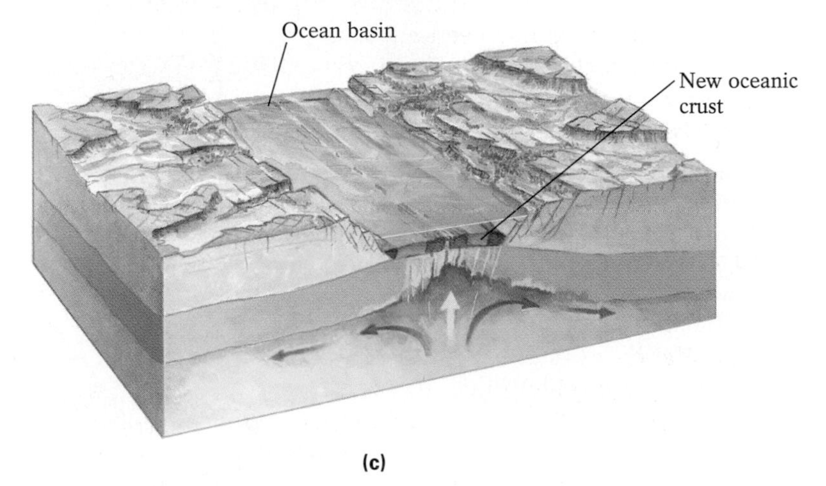

(c)

Figure 1-9 Plate rifting and divergence. When currents in the underlying asthenosphere pull one of the Earth's plates in opposite directions **(a),** the plate is stressed and eventually rifts. **(b)** As the plate fragments continue to move (diverge) farther from one another, molten rock from the mantle rises into the gap and solidifies along the edges of the plates **(c),** forming new oceanic crust that is eventually covered with water to form a new ocean basin.

4. The interiors of plates are relatively quiet geologically, with little volcanic activity and far fewer and usually milder earthquakes than occur at plate boundaries.

Figure 1-8 shows the Earth's seven large, or major, plates and a number of its smaller ones. Note that the continents themselves are not plates, but instead are usually parts of composite plates that contain both continental and oceanic portions. As these plates move, everything on them—including continents and oceans—moves with them. For example, as the North American plate moves westward away from the Eurasian plate, both the continent of North America and the western half of the Atlantic Ocean are moving farther away from Europe.

Although the fact that entire continents and ocean basins are in motion is rather astounding, the rate at which they move is not; it is comparable to the rate of growth of your fingernails—only a few centimeters per year. If Columbus were crossing the Atlantic Ocean today, 500 years after his famous voyage, he would have to sail only an extra 30 to 50 meters (100–160 feet) to reach shore. Over the vast course of geologic time, however, the rate of plate movement has been sufficient to open and close the Atlantic Ocean several times.

Plate Movements and Boundaries

The Earth's plates move relative to each other in several ways, and plate boundaries are categorized according to which type of movement they demonstrate. Three major types of boundaries exist: divergent plate boundaries, where plates move apart; convergent plate boundaries, where plates move together; and transform plate boundaries, where plates move past one another in opposite directions.

Rifting and Divergent Plate Boundaries Within plate interiors, comparatively little geological activity takes place. Nevertheless, plate interiors may become geologically active if slow-flowing currents in the Earth's asthenosphere generate a pulling-apart motion that tears a preexisting plate into two or more smaller plates. This process, discussed in detail in Chapter 11, is known as **rifting.** The Great Rift Valley of East Africa, where the African plate has been coming apart, is a prime example of early rifting.

Once a plate has been rifted, the resulting smaller plates may continue to separate from one another, a type of plate motion known as **divergence** (Fig. 1-9). Divergence typically

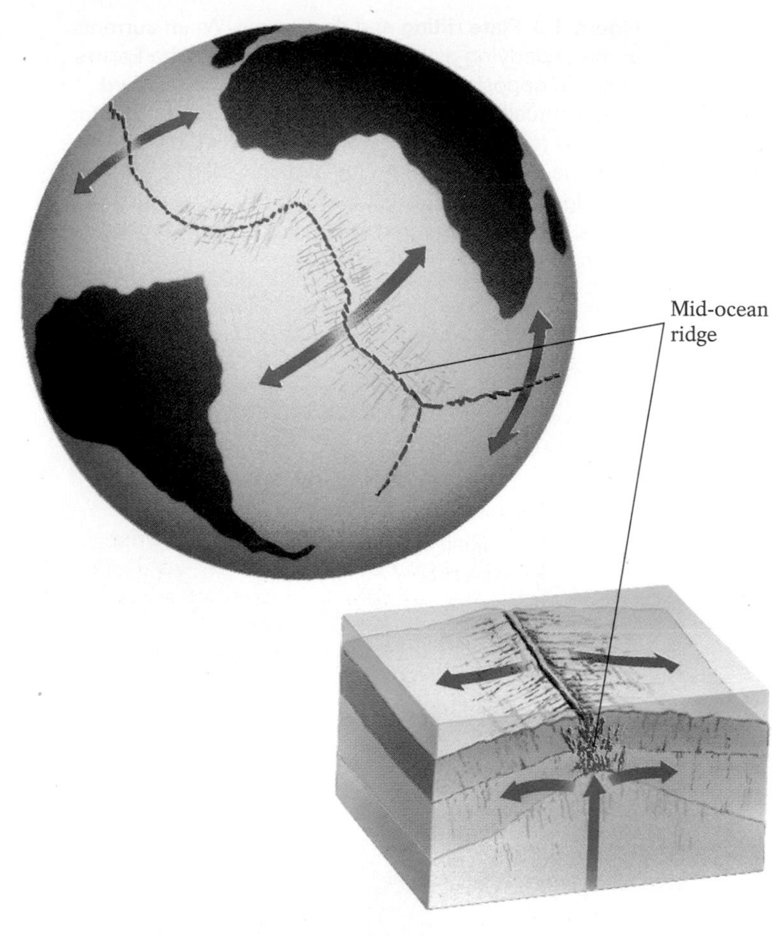

Figure 1-10 **The mid-ocean ridge.** This region, where molten rock from the Earth's interior erupts and cools to become new lithosphere, forms the Earth's longest mountain range, extending for more than 64,000 kilometers (40,000 miles).

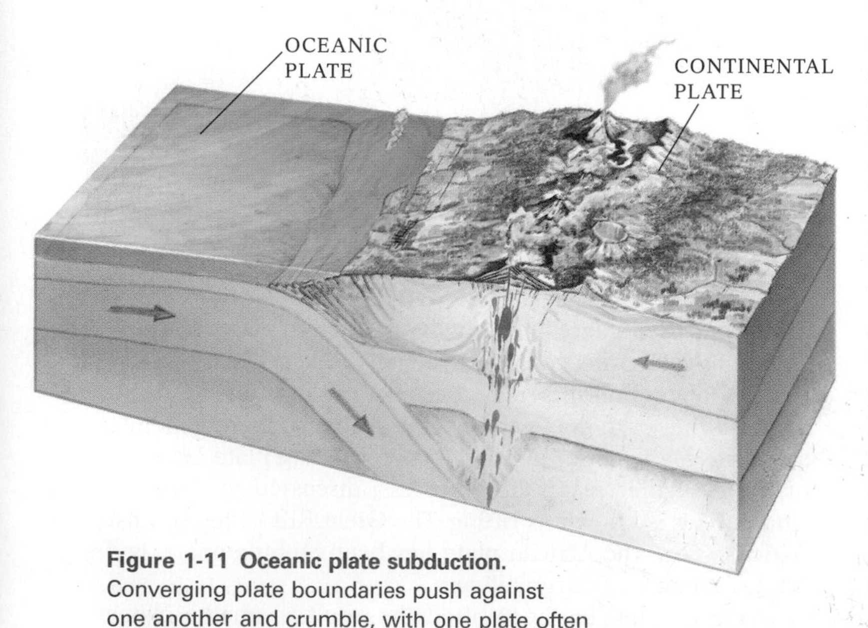

Figure 1-11 **Oceanic plate subduction.** Converging plate boundaries push against one another and crumble, with one plate often sinking, or subducting, below the other.

proceeds at a rate of about 1 to 10 centimeters (0.5–4 inches) per year, as molten rock rises into the thousands of fractures between the rifted plates, cools and solidifies, and becomes attached to the edges of the rifted plates. Meanwhile divergence continues, further separating the older rifted segments and eventually forming an ocean basin. The new ocean basin fills with seawater as further rifting opens new connections to other oceans.

Throughout the period of divergence, erupting molten rock expands the ocean basins by creating new oceanic crust. The center of this volcanic activity is the *mid-ocean ridge,* a continuous chain of submarine mountains that meanders around the globe like the stitches on a baseball (Fig. 1-10). The process of plate growth at mid-ocean ridges is known as **sea-floor spreading.** If the young Red Sea between the African and Arabian plates (see Fig. 1-8) continues to grow at its present rate, it may eventually become a full-blown ocean like the Atlantic or Pacific.

Plate Convergence and Subduction Boundaries Plate **convergence** occurs when plates move toward each other. Convergence may involve two continental plates, two oceanic plates, or one of each type. When two oceanic plates converge, the denser of the two plates sinks beneath the other and is reabsorbed into the Earth's interior, a process known as **subduction.** Because plates of oceanic lithosphere are always denser than those of continental lithosphere, when these two types of plates converge the oceanic plate always subducts (Fig. 1-11).

Subduction itself produces a number of other geological phenomena. For example, friction between the two plates often produces earthquakes. Partial melting of the subducting plate as it reaches the asthenosphere creates a magma that may rise to form volcanoes. This phenomenon is observed in the northern Pacific, where the Pacific plate descends beneath the oceanic edges of the North American plate to form the earthquake-wracked volcanic Aleutian Islands of Alaska. The 1996 eruption of Mount Pavlov in Alaska provided evidence of ongoing subduction in this region.

Continental plates are generally too buoyant to subduct into the denser underlying mantle. Thus, when two continental plates collide, neither plate can subduct completely, although both plates' edges may be temporarily dragged down to depths of perhaps 200 kilometers (120 miles) before being thrown back toward the Earth's surface. Instead of subducting, colliding continental plates become welded together, pushing the colliding edges upward and forming a much larger single plate. Such convergence of continental plates, called **continental collision** (Fig. 1-12), has created many of the Earth's largest mountain ranges. Northern Africa's Atlas Mountains and southern Europe's Alps were formed by past collisions of the African and Eurasian plates, the Himalayas from the ongoing collision of the Indian and Eurasian plates, and North America's Appalachians from a three-way collision of the African, Eurasian,

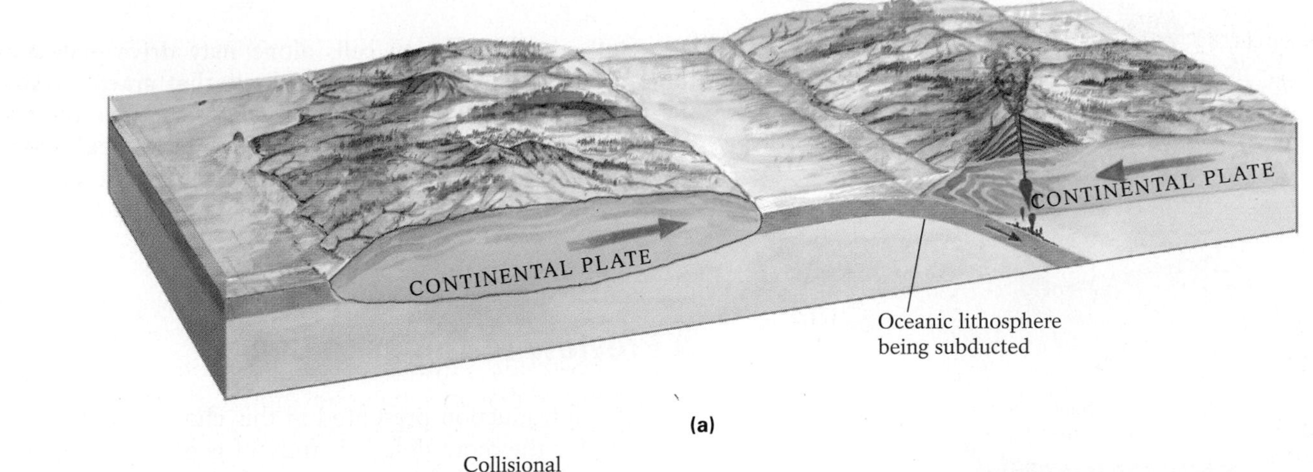

(a)

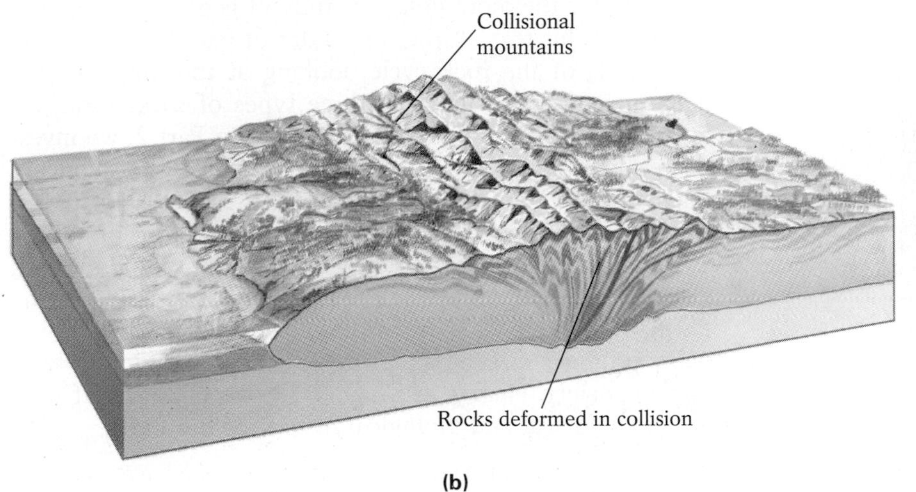

(b)

Figure 1-12 Continental collision. **(a)** Two continental plates converge as the oceanic lithosphere between them becomes subducted. **(b)** With the intervening oceanic lithosphere completely subducted, the two continental plates collide and are uplifted, because neither is dense enough to subduct. The collision results in a mountain range composed of highly deformed rocks. Note the thickened plate at the point of collision.

and North American plates that took place between about 400 and 250 million years ago.

Transform Motion and Transform Plate Boundaries The third major type of plate boundary occurs where two plates, either oceanic or continental, move past one another in opposite directions, a process known as **transform motion** (Fig. 1-13). Because great friction results as the moving plates grind past each other, these boundaries produce earthquakes, and any community situated near such a boundary may be periodically devastated. San Francisco and Los Angeles, for example, are both located within the San Andreas transform zone between the North American and Pacific plates.

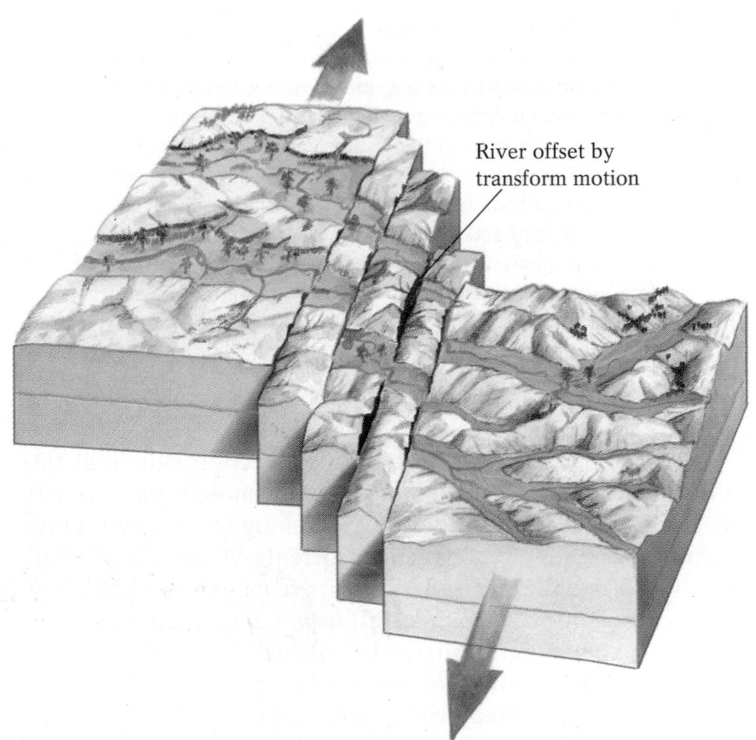

Figure 1-13 Transform motion. When the Earth's plates move past one another in opposite directions, friction builds up at their edges but the plates are neither uplifted nor subducted.

The Driving Force Behind Plate Motion

Geologists believe that heat-driven currents within the Earth's mantle are principally responsible for plate movements. These currents, known as **convection cells,** develop when portions of the asthenosphere are heated, become less dense than the surrounding material, and rise toward the surface. When the moving material encounters the solid lithosphere, it spreads

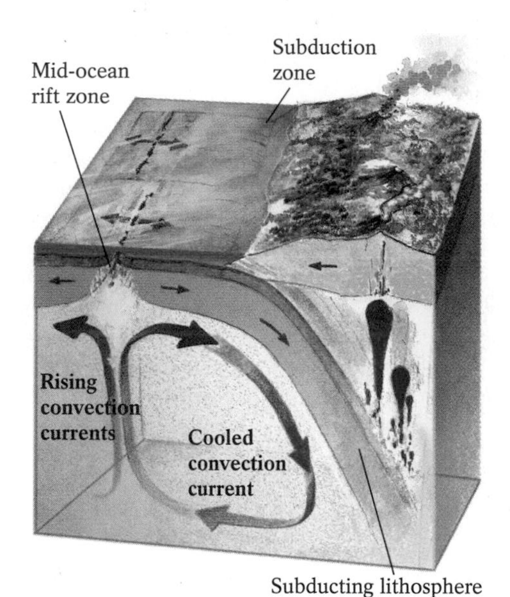

Figure 1-14 Convection cells and plate motion. Heat within the Earth's mantle results in rising ("convecting") currents of warm mantle material, which drag the lighter lithospheric plates along with them as they flow beneath the Earth's surface. As rising mantle material spreads beneath the lithospheric plates, it cools, becomes denser, and sinks back to the deeper interior, where it is reheated to rise again. Such a cycle, known as a convection cell, may be the principal driving mechanism of plate tectonics.

laterally beneath it and drags the lithospheric plates along in a conveyor-belt–like fashion (Fig. 1-14).

In some places, neighboring convection cells pull the lithosphere in opposite directions with enough force to rift it; the rifted plates are then carried along in opposite directions by the slowly convecting currents. A relatively small amount of mantle-derived material escapes to the surface at such divergent plate boundaries, forming a mid-ocean ridge; sea-floor spreading occurs as the plates continue to diverge and mantle material rises to the surface and cools to form new lithosphere. Meanwhile, the leading edges of the diverging plates eventually encounter the edges of other plates at convergent zones. Here, the plate edges are either welded together in continental collisions or subducted back into the asthenosphere to reenter the cycle.

Although convection cells alone may drive plate tectonic processes, some geologists believe that gravity assists the process by literally dragging the plates into the interior of the Earth at subduction zones. Thus both heat-driven divergence and gravity-driven subduction may work together to cycle the Earth's plates.

A Preview of Things to Come

The brief introduction presented in this chapter should prepare you for the more detailed study of geology throughout the rest of this text. In the remainder of Part 1, we examine the details of the rock cycle, looking at the minerals that make up rocks and how the three types of rocks form. We also take a closer look at geologic time. In Part 2, we investigate plate tectonics in depth and discuss how this concept relates to earthquakes and to mountain building. In Part 3, we explore the processes that transform the landscapes of the Earth, looking at the power of gravity, streams, groundwater, glaciers, wind, and waves as they sculpt the surface of the Earth. In Chapter 18, we examine Earth's resources—the many products of geologic processes that have found use in human society. Finally, in Chapter 19, we take a brief look at the geological and biological history of the Earth.

Chapter Summary

Geology is the scientific study of the Earth. In applying the **scientific method,** geologists systematically collect data derived from experiments and observations. They analyze and interpret their findings and develop **hypotheses** to explain how the forces of nature work. A hypothesis that is consistently supported by further study and investigation may be elevated to the status of a widely accepted explanation, or **theory.** A theory that withstands rigorous testing over a long period of time may be declared a **scientific law.**

The hypothesis of **catastrophism,** which was popular until the mid-eighteenth century, held that the Earth had been formed through a series of immense worldwide upheavals. **Uniformitarianism,** which by 1830 had replaced catastrophism, suggests that the Earth has been formed predominantly by slow, gradual, small-scale processes that still operate today.

The universe is believed to have begun with the Big Bang about 12 billion years ago. The Sun, which is a star, formed nearly 5 billion years ago from the collapse of a gas cloud, the center of which heated up as particles drawn inward by gravity collided and produced nuclear reactions. As the outer region of the gas cloud cooled, the Earth and other planets developed (with the Earth being formed about 4.6 billion years ago) by accretion of colliding bits of matter. During the Earth's first several million years, the impact of these accreted masses warmed its interior until the accumulated heat was sufficient to melt most of the planet's constituents. At this time, **differentiation** took place, in which the Earth's densest elements (primarily iron and nickel) sank toward its interior while its lightest elements moved upward to its surface. Today, the Earth has three concentric layers of different densities: a thin, least dense outer layer, called the **crust;** a thick, more dense underlying layer, called the **mantle;** and a much smaller **core,** which is the densest of Earth's layers. The Earth's **lithosphere,** a composite layer made up of the crust and the outermost segment of the mantle, is solid and brittle; below it lies the flowing, heat-softened rock of the **asthenosphere.** Cosmologists hypothesize that the Earth's Moon formed roughly 4.55 billion years ago as a result of a cataclysmic collision between the Earth and a Mars-sized planetesimal.

Rocks, which are defined as naturally occurring aggregates of one or more minerals, are categorized according to the way in which they form. The three basic rock groups are as follows: **igneous rocks,** which solidify from molten material; **sedimentary rocks,** which are compacted and cemented aggregates of fragments of preexisting rocks of any type; and **metamorphic rocks,** which form from any type of rock when its chemical composition is altered by heat, pressure, or chemical reactions in the Earth's interior. The continual transformation of the Earth's rocks from one type into another over time is called the **rock cycle.**

The modern theory of **plate tectonics** states that the Earth's lithosphere is composed of seven major and a dozen or more minor plates, which move in response to the flow of the asthenosphere below them. The plates move relative to each other in three ways: away from one another, by **divergence;** toward one another, by **convergence;** or past one another in opposite directions, by **transform motion.** Most large-scale geological activity, such as earthquakes and volcanoes, occurs at plate boundaries.

Divergence is preceded by **rifting,** a process in which a large preexisting plate is torn into two or more smaller plates. New oceanic lithosphere forms by the process of **sea-floor spreading,** in which molten rock from the Earth's mantle wells upward, cools, and solidifies between rifting plates. Convergence involving either two oceanic plates or one oceanic plate and one continental plate results in **subduction,** in which the denser of the two plates sinks below the other and is reabsorbed into the Earth's mantle. A **continental collision** occurs when both converging plates are continental, in which case neither is dense enough to subduct and the plates' edges are welded together and uplifted into mountains by the pressure of the collision.

The primary force that drives tectonic plates appears to be heat-driven **convection cells** within the Earth's mantle. Near the surface, these convection cells drag the lithospheric plates along, causing plate movement.

Key Terms

geology (p. 4)
scientific method (p. 5)
hypothesis (p. 5)
theory (p. 5)
scientific law (p. 6)
catastrophism (p. 6)
uniformitarianism (p. 6)
differentiation (p. 9)
crust (p. 9)
mantle (p. 9)
lithosphere (p. 9)
asthenosphere (p. 9)
core (p. 9)
rock (p. 10)
igneous rocks (p. 10)
sedimentary rocks (p. 10)
metamorphic rocks (p. 10)
rock cycle (p. 10)
plate tectonics (p. 12)
rifting (p. 13)
divergence (p. 13)
sea-floor spreading (p. 14)
convergence (p. 14)
subduction (p. 14)
continental collision (p. 14)
transform motion (p. 15)
convection cells (p. 15)

Questions for Review

1. Briefly explain the difference between a scientific hypothesis, a scientific theory, and a scientific law.

2. Contrast the principles of catastrophism and uniformitarianism.

3. How did the originally homogeneous Earth become differentiated into concentric layers?

4. Draw a simple sketch of the major layers that make up the Earth's interior. Which of these layers form the Earth's plates?

5. Describe the three major types of rocks in the Earth's rock cycle.

6. List the four basic concepts of the theory of plate tectonics.

7. Draw simple sketches of divergent plate boundaries, two kinds of convergent plate boundaries, and transform plate boundaries.

8. At which type of plate boundary does oceanic lithosphere form? At which type of plate boundary is oceanic lithosphere consumed?

9. Why are there so few earthquakes in Minneapolis and Indianapolis?

10. Draw a simple sketch of a convection cell and explain how it might drive plate movement.

For Further Thought

1. When geologists find ancient glacial deposits in equatorial Africa, they usually interpret them as polar deposits that have drifted from a cold place to a warm place. Formulate another hypothesis to explain this phenomenon.

2. Why is there no current volcanic activity along the eastern coast of North America?

3. Describe how plate tectonic activity might affect the rock cycle.

4. Find the Ural Mountains on a map of Eastern Europe. Briefly explain how they might have formed.

5. As recently as 5 million years ago, South America and North America were completely separated, unattached by Central America. Using Figure 1-8, speculate about how the Central American connection that links the Western Hemisphere might have formed.

2 Minerals

Many cultures have long valued minerals for their sheer visual appeal—for their stunning colors or lusters, or their often-perfect symmetry (Fig. 2-1). But the importance of minerals is hardly just aesthetic and is not restricted to the "pretty" specimens found in museums or jewelers' showcases. From the simple flint hand scrapers made by our ancestors hundreds of thousands of years ago to the quartz crystals in modern clocks and watches, minerals have played a fundamental role in human life (Fig. 2-2). Every day we use a vast array of minerals, all derived from the Earth, in a remarkable number of ways. For example, a common absorbent mineral called talc (found in talcum powder) is used to dry and soothe our skin. Other minerals provide the sulfur used to manufacture fertilizers, paints, dyes, detergents, explosives, synthetic fibers, and books of matches, and the fluorine that helps refrigerate food and cool homes and offices. Aluminum,

Figure 2-1 Crystals of quartz. These samples illustrate the beauty and symmetry for which many minerals are valued.

(a)

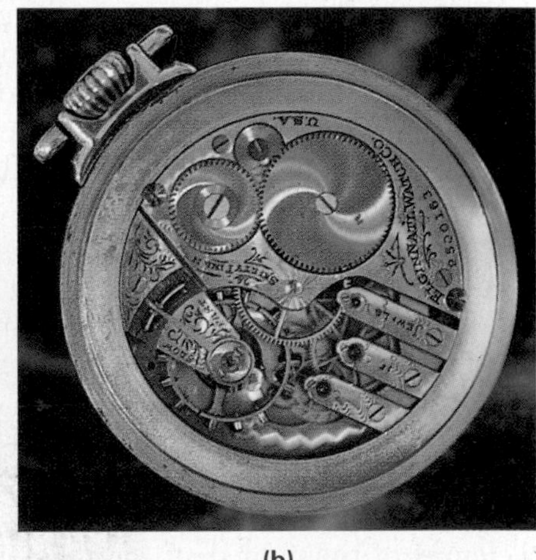

(b)

Figure 2-2 Uses of minerals. Minerals have been providing us with essential tools for hundreds of thousands of years. **(a)** This flint hand scraper was used by Native Americans more than a thousand years ago, probably to scrape flesh and hair from animal skins. **(b)** This elegant personal timepiece runs with great accuracy thanks to the constant vibration of a quartz crystal.

commonly derived from the mineral bauxite, is an ideal component for space shuttles, garden furniture, window frames, and beer and soft-drink cans because of its lightness, strength, and resistance to corrosion. When you're laid low by a common intestinal problem, you may run for a spoonful of Kaopectate, a remedy whose active ingredient is the mineral kaolinite.

Many of the nutrients our bodies need come from minerals in the soil that become incorporated into the fruits and vegetables we eat. These minerals include calcium, phosphorus, and fluorine, which give our bones and teeth their hardness, sodium and potassium, which regulate our blood pressure, and iron, which helps our blood carry life-sustaining oxygen.

In this chapter, we examine what minerals are and how they are formed, and discuss methods of identifying the different kinds of minerals. We also look at the distinctive characteristics of several important minerals and mineral groups.

Figure 2-3 Granite, an igneous rock. One can clearly see the individual mineral components in this piece of granite.

The Chemistry of Minerals

Minerals are naturally occurring inorganic solids consisting of one or more chemical elements in specific proportions, whose atoms are arranged in a systematic internal pattern. Diamond, calcite, and quartz are examples of minerals. Because minerals are *naturally* occurring solids, the thousands of synthetic compounds produced in laboratories do not qualify as minerals. Because minerals are *inorganic,* coal, which is composed of heated and compressed remnants of plants, is not considered a mineral. Because minerals have a *systematic internal organization,* substances whose atoms do not follow such a pattern, such as the gemstone opal, are not considered true minerals. **Rocks** are naturally occurring aggregates, or combinations, of one or more minerals, with each mineral retaining its own discrete characteristics. For example, the rock granite contains minerals such as quartz, plagioclase, and hornblende (Fig. 2-3).

Geology students sometimes wonder why so much discussion about minerals relates to their chemistry. The principal reason is that all minerals are composed of combinations of chemical elements, and these chemical structures determine the minerals' distinctive characteristics. These characteristics, in turn, determine their uses and value to society. To know why diamonds are hard, why gold can be pounded into wafer-thin leaves, and why we build skyscrapers with skeletons of titanium steel, we must understand the chemical makeup of minerals.

Minerals are composed of one or more elements in specific proportions. An **element** is a form of matter that cannot be broken down into a simpler form by heating, cooling, or reacting with other chemical elements. Aluminum and oxygen are two common elements. **Atoms** (from the Greek *atomos,* meaning "indivisible") are the smallest particles of an element that retain all its chemical characteristics. All atoms of a given element are identical in certain fundamental ways and differ in these ways from the atoms of every other element.

More than 112 elements are known, of which 92 occur naturally and 20 are laboratory creations. Every element can be represented by its chemical symbol, a one- or two-letter abbreviation. These symbols usually consist of the first letter or letters of the English or Latin name of the element, such as O for oxygen, Al for aluminum, and Na (from the Latin *natrium*) for sodium. Chemists have arranged all the known elements into the Periodic Table of Elements (Fig. 2-4).

Atoms of two or more elements may combine in specific proportions to form chemical **compounds.** The fixed proportions of atoms that make up a compound are expressed by combinations of symbols in a *chemical formula;* for example, the silicon and oxygen compound that makes up the mineral quartz has the chemical formula SiO_2. The subscript numerals denote the ratio of atoms of each element in the chemical compound; thus its formula reveals that quartz contains two atoms of oxygen for every atom of silicon.

The Structure of Atoms

An atom is incredibly small, approximately 0.00000001 centimeter (one hundred-millionth of a centimeter) in diameter. This line of type would contain about 800,000,000 atoms laid side by side. As small as an atom is, it consists of even smaller particles: **protons** and **neutrons,** located in the central region, or **nucleus,** of an atom, plus **electrons,** moving

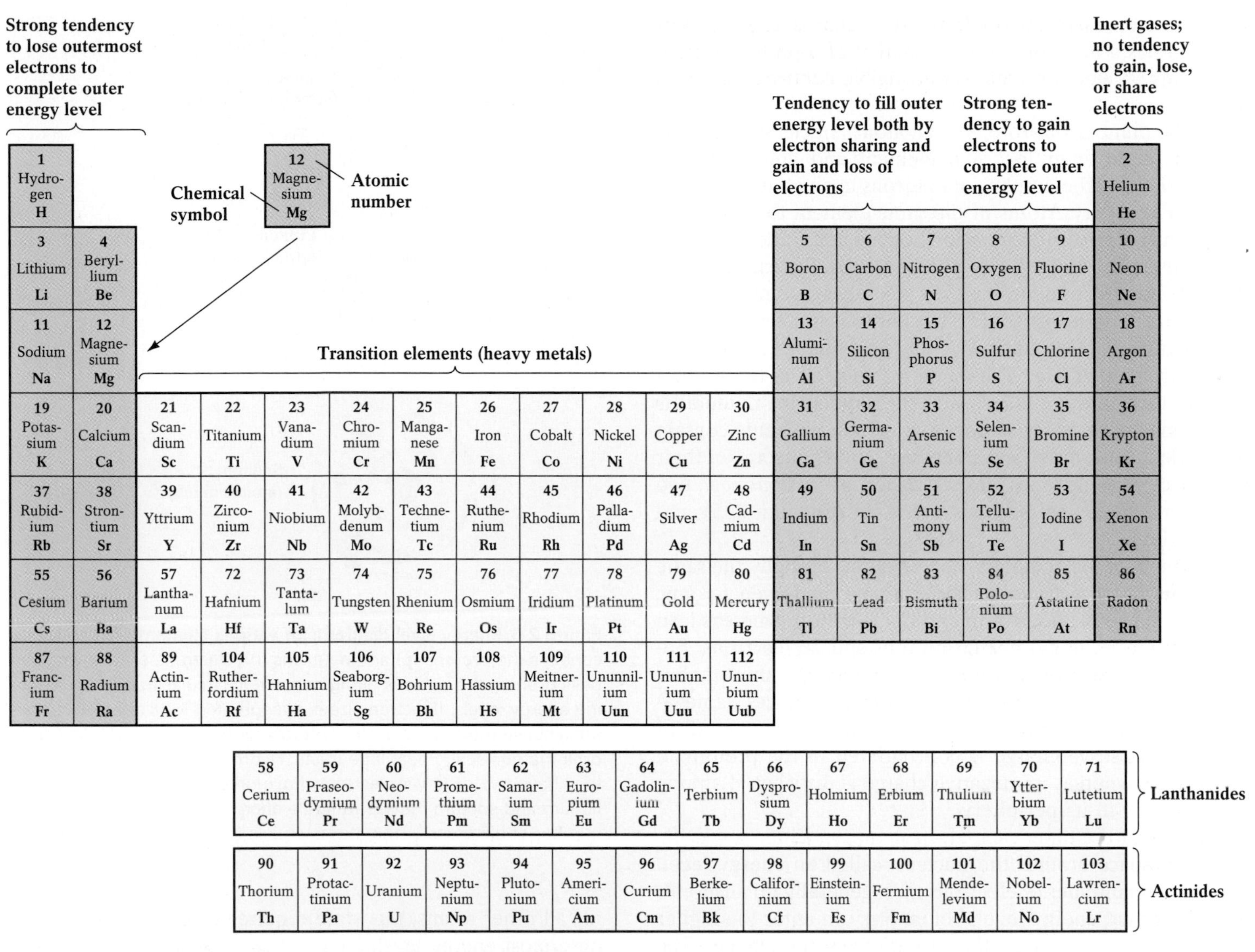

Figure 2-4 The Periodic Table of Elements. The periodic table groups all elements by similarities in their atomic structures, which result, in turn, in similarities in their chemical properties.

about outside the nucleus. Figure 2-5 depicts a simplified model of an atom.

Each proton carries a single positive charge, expressed as +1, and has a mass of 1.67×10^{-24} gram, which for convenience is referred to as an *atomic mass unit* (AMU) of 1. Neutrons are nearly identical to protons in size and mass, but, as their name suggests, they have no charge—they are neutral. Neutrons do, however, contribute to the **atomic mass** of the atom, which is the total mass of all protons and neutrons within an atom's nucleus. Thus an atom containing one proton and one neutron has an atomic mass of 2 AMU. The third type of atomic particle, the electron, stays in motion around the nucleus at speeds so great that if one were orbiting the Earth, it would do so in less than a single second.

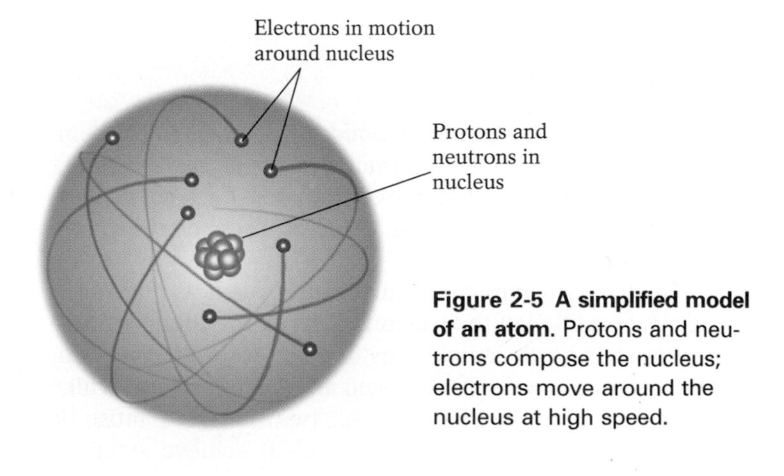

Figure 2-5 A simplified model of an atom. Protons and neutrons compose the nucleus; electrons move around the nucleus at high speed.

Each electron carries a single negative charge, expressed as −1, and has a mass of about 1/1836 that of a proton or neutron; thus an electron makes a negligible contribution to an atom's mass.

The number of protons in an atom's nucleus, its **atomic number,** is constant for each element and determines an atom's identity. The number of neutrons in an atom's nucleus, however, can vary. Atoms of the same element that have different numbers of neutrons in their nucleus, and therefore different atomic masses, are called **isotopes.** For example, the element oxygen (atomic number = 8) always contains eight protons in its nucleus, but it has three isotopes: $^{16}_{8}O$, with eight neutrons in its nucleus; $^{17}_{8}O$, with nine neutrons; and $^{18}_{8}O$, with ten neutrons. The subscript numeral in these notations is the atomic number, and the superscript numeral is the atomic mass. Some isotopes of certain elements contain nuclei that break down spontaneously and emit some of their particles. Such isotopes are described as *radioactive.* Two common radioactive isotopes are $^{235}_{92}U$ (uranium-235) and $^{14}_{6}C$ (carbon-14).

The number of electrons in an atom is usually the same as the number of protons. For example, hydrogen (atomic number = 1) has one proton and one electron, whereas iron (atomic number = 26) has 26 protons and 26 electrons. Because the number of an atom's protons equals the number of its electrons, its positive charge exactly balances its negative charge and the atom therefore has no net charge. All of an atom's positive charge is concentrated in the protons in its nucleus, whereas its negative charge is distributed among the electrons in its periphery.

Most of the time, each electron moves within a specific region of space around the nucleus, called an **energy level.** A maximum number of electrons must generally fill an atom's lowest, or first, energy level, before any can enter the higher energy levels, which are more distant from the nucleus. The lowest energy level in any atom always has a maximum capacity of two electrons. The second energy level can hold a maximum of eight electrons, and succeeding energy levels can each hold eight or more electrons. Figure 2-6 provides a schematic of the number and energy-level positions of some atoms' electrons.

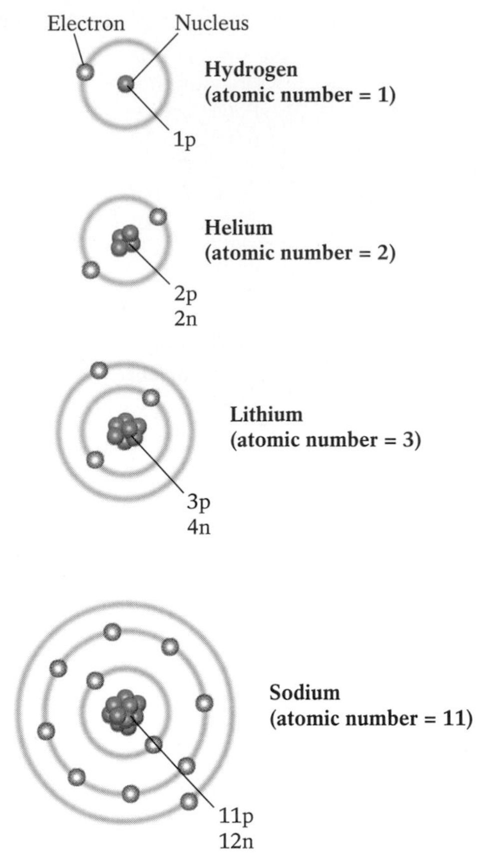

Figure 2-6 Energy-level diagrams of various elements. The nucleus contains the protons (p) and neutrons (n); electrons are shown as balls orbiting the nucleus along concentric circular tracks representing energy levels. (Electron size is exaggerated for clarity; electrons are actually much smaller than protons and neutrons and do not orbit the nucleus in neat little tracts. Hydrogen and helium, because they have two or fewer electrons, have only one energy level; lithium and sodium require multiple energy levels. Note how the number of electrons in the atoms equals the number of protons.

Bonding of Atoms

Atoms frequently combine, or **bond,** to form chemical compounds. Two key factors determine which atoms will unite with which other atoms to form compounds: Each atom tends toward chemical stability, and an electrically neutral compound is more stable.

An atom achieves chemical stability when its outermost energy level is filled with electrons. Thus atoms bond by transferring or sharing electrons so as to attain full outermost energy levels. In the case of hydrogen and helium atoms, which have only the lowest energy level, two electrons must be present in the outermost energy level to achieve stability; for all other atoms, this state requires eight electrons in the outermost energy level.

Atoms of some elements tend to lose their outer electrons, whereas others tend to gain electrons, depending largely on the number of electrons in their outermost energy levels. Elements with one or two electrons in their outermost energy levels, for example, have a strong tendency to give up those electrons. In contrast, elements with six or seven electrons in their outermost energy levels tend to acquire electrons. Most other elements tend to share electrons with other atoms instead of transferring or receiving them. Elements whose outer energy levels are already full are very chemically stable, or *inert*—they do not lose, gain, or share electrons, and they are unlikely to bond with other atoms.

Because of the diversity of electron configurations among elements, various types of bonding are possible. The atoms that make up the vast majority of the Earth's minerals are most often linked by ionic bonding, covalent bonding, or metallic bonding. Another type of bonding—intermolecular bonding—affects minerals principally because of the way it bonds compounds that react with minerals.

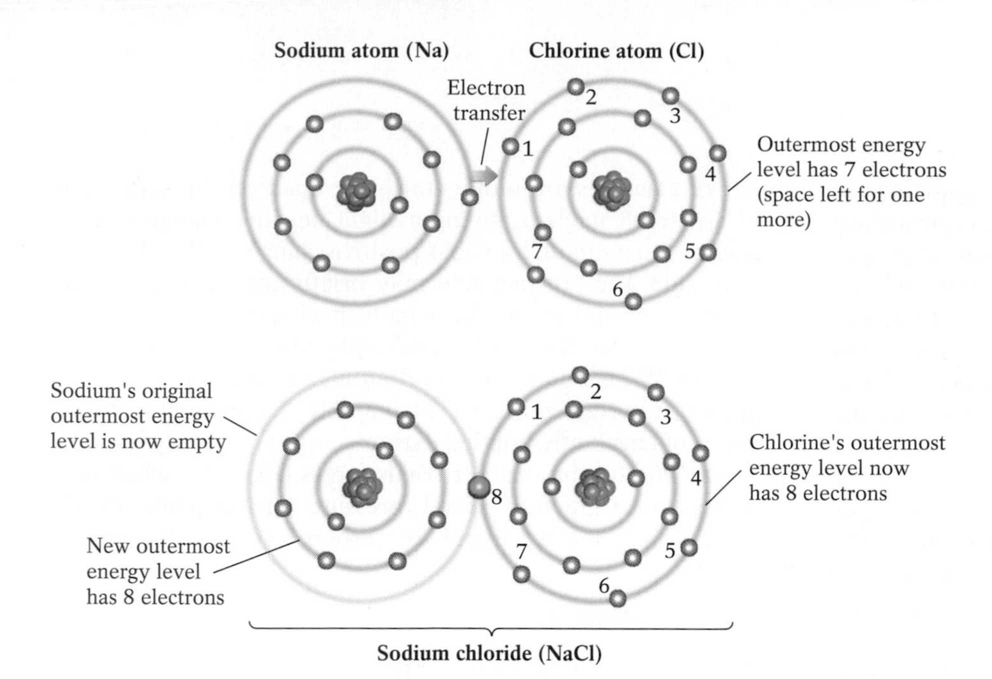

Figure 2-7 Ionic bonding of sodium (Na) and chlorine (Cl). When a sodium atom (with one electron in its outer energy level) donates its outermost electron to a chlorine atom (with seven electrons in its outer energy level), the outer energy level in the new configuration of each atom has eight electrons. The two resulting ions (Na^+ and Cl^-) unite to form sodium chloride (NaCl), a neutral ionic compound.

Ionic Bonding An atom does not change its identity when it loses or gains an electron, but it does lose its electrical neutrality, thereby becoming a positively or negatively charged **ion.** For example, when a chlorine atom gains a single electron, it becomes chemically stable, but it also becomes a negatively charged ion, symbolized by Cl^-. When a potassium atom loses one electron and a calcium atom loses two electrons, they become positively charged ions, symbolized by K^+ and Ca^{2+}, respectively.

When an atom with a strong tendency to lose electrons, such as sodium, comes in contact with an atom with a strong tendency to gain electrons, such as chlorine, electrons are generally transferred so that each atom achieves the chemical stability of a full outer energy level (Fig. 2-7). Atoms that lose one or more electrons become positively charged ions; atoms that gain one or more electrons become negatively charged ions. These oppositely charged ions then attract each other to form **ionic bonds.** The result is an electrically neutral, chemically stable compound.

The physical and chemical properties of compounds differ from those of their component elements. Pure sodium, for instance, is a soft, silvery metal that reacts vigorously when mixed with water and may even burst into flame; chlorine usually occurs as Cl_2, a green poisonous gas that was used as a weapon during World War I. Combining the two results in sodium chloride, which neither ignites nor poisons—it is the white crystalline mineral halite (table salt), a substance that regulates some of the biochemical processes essential to all life.

Covalent Bonding Atoms whose outer energy levels are approximately half full (containing three, four, or five electrons) tend to achieve chemical stability by sharing electrons with other atoms. In such a case, both atoms fill their outer energy levels with the shared electrons rather than transferring electrons from one to the other. Sharing electrons produces a **covalent bond,** in which the outer energy levels of the atoms overlap. Covalent bonds are generally the strongest type of bond. In some cases, two or more atoms of a single element may bond covalently with each other. For example, overlapping carbon atoms bond covalently in diamond, an all-carbon mineral (Fig. 2-8).

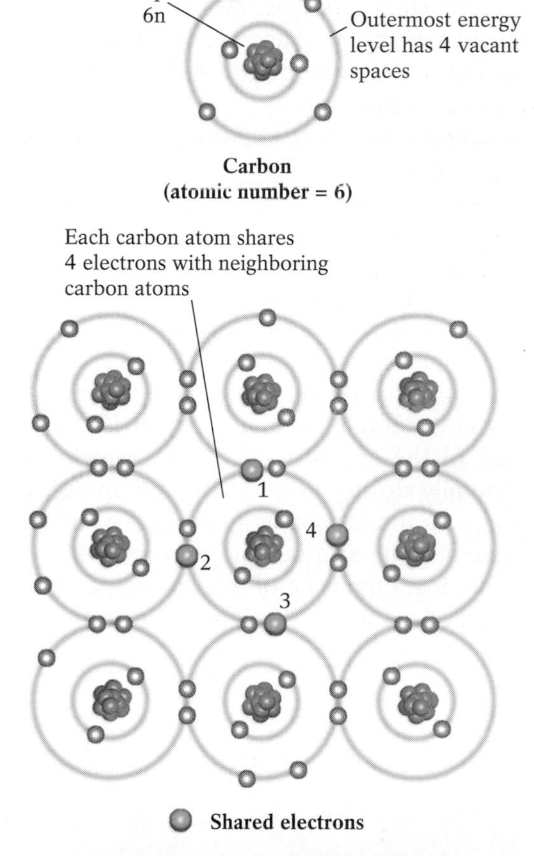

Figure 2-8 Covalent bonding in diamond, a mineral that consists entirely of the element carbon. Each carbon atom in diamond is bonded covalently to four neighboring carbon atoms; the great strength of these bonds accounts for the fact that diamond is the hardest known substance on Earth.

Metallic Bonding The atoms of some electron-donating elements tend to pack closely together, with each typically surrounded by either eight or twelve others. This arrangement produces a cloud of electrons that roam independently among the positively charged nuclei, unattached to any specific nucleus. In this phenomenon known as **metallic bonding,** the roaming electrons in the metallically bonded substances allow the metals to function as efficient conductors of electricity.

Intermolecular Bonding Certain compounds (but not minerals) exist as *molecules,* stable groups of bonded atoms that are the smallest particles identifiable as compounds. For example, one molecule of water, H_2O, consists of two atoms of hydrogen bonded to one atom of oxygen. Molecules are often attached weakly to other molecules by *intermolecular bonds.* Intermolecular bonding results from the relatively weak positive or negative charges that develop at different locations within a molecule due to the uneven distribution of its moving electrons.

For geologists, the most important type of intermolecular bonding involves water. The uneven distribution of electrons in a water molecule causes it to have a slightly negative side (near its oxygen atom) and a slightly positive side (near its hydrogen atoms). These charged regions in water molecules can attract oppositely charged ions from the surface of some minerals, forming weak **hydrogen bonds** with them. The combined effect of many such bonds can break the internal bonds of some minerals; as a consequence, table salt and many other minerals dissolve in water (see Chapter 5).

Because of the circumstances under which they form, minerals generally do not occur as discrete molecules. Nevertheless, because the groups of atoms that compose minerals have uneven charge distributions, they are subject to intermolecular bonding. One type of intermolecular bond found in minerals, called a **van der Waals bond,** forms when a number of electrons are momentarily grouped on the same side of an atom's nucleus, giving a slight negative charge to that side of the atom and a slight positive charge to the electron-poor side. The positive side may briefly attract electrons of neighboring atoms, and the negatively charged side may fleetingly attract the nuclei of neighboring atoms. Although weak, van der Waals bonds can bond atoms or layers of atoms together in certain minerals. In graphite, for example, sturdy layers of covalently bonded carbon atoms are weakly bonded to one another by van der Waals forces (Fig. 2-9); when you write with a graphite pencil, pressure on the point breaks these bonds, leaving a trail of carbon layers on the paper.

Mineral Structure

When a mineral grows unrestricted in an open space, it develops into a regular geometric shape known as a **crystal** (from the Greek *kyros,* meaning "ice"). This shape is the external expression of the mineral's internal **crystal structure,** the orderly arrangement of its ions or atoms into a latticework of repeated three-dimensional units. As noted earlier, this systematic internal organization is a defining characteristic of all minerals.

Because the internal structure of a given mineral is always the same, the shape of its crystals will be the same in every well-formed, unbroken sample of the mineral. A perfect quartz crystal from Herkimer, New York, for instance, is identical to a perfect quartz crystal from Hot Springs, Arkansas. Most minerals form in restricted growing spaces, however, which prevents them from developing into perfect crystals. As a result, although all samples of a given mineral possess the same internal crystal structure, they may not take the same shape externally.

Sometimes molten rock may cool too rapidly for its atoms to form even a semblance of an orderly arrangement. The resulting solids, called **mineraloids,** lack a specific crystal

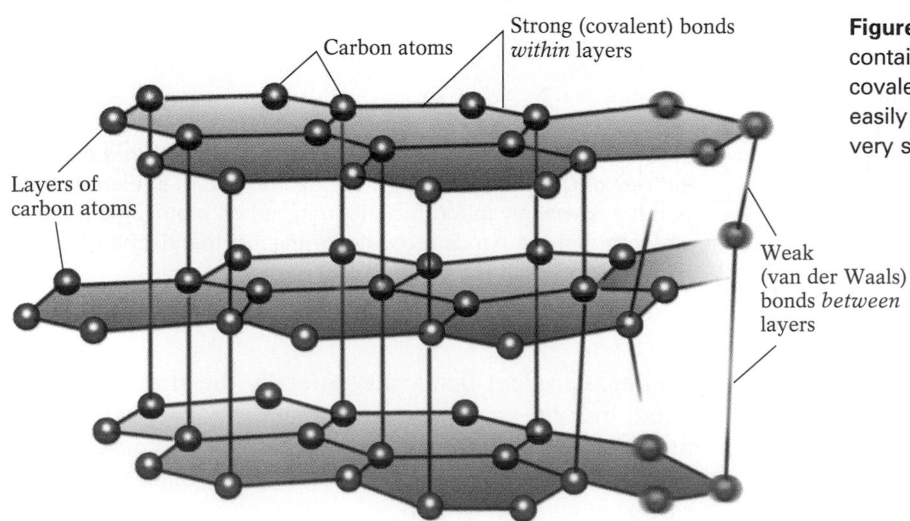

Figure 2-9 Graphite, another form of carbon. Graphite contains weak van der Waals bonds between layers of covalently bonded carbon atoms. Because graphite is easily broken at the sites of these weak bonds, it is a very soft mineral.

structure and thus do not qualify as true minerals. The volcanic rock obsidian, a type of natural glass, is an example of a mineraloid.

Determinants of Mineral Formation

The kinds of minerals that form in a particular time and place depend on the relative abundances of the available elements, the relative sizes and other characteristics of those elements' atoms and ions, and the temperature and pressure at the time of formation.

Only eight of the 92 naturally occurring elements in the Earth's continental crust are relatively abundant, with oxygen and silicon dominating (Table 2-1). Thus most minerals in the crust are oxygen- and silicon-based compounds. Most minerals in the upper mantle are oxygen-silicon-iron-magnesium–based compounds.

Table 2-1 The Most Abundant Elements in the Earth's Continental Crust

Element	Proportion of Crust's Weight (%)
Oxygen (O)	45.20
Silicon (Si)	27.20
Aluminum (Al)	8.00
Iron (Fe)	5.80
Calcium (Ca)	5.06
Magnesium (Mg)	2.77
Sodium (Na)	2.32
Potassium (K)	1.68
	98.03
Other elements	1.97
Total	100.00

In addition to the relative abundance of available elements, mineral formation depends on how readily these elements interact with one another to form a crystalline structure: Given two elements of equal abundance, the element that will contribute more readily to mineral formation will be the one that "fits" better with the other elements present. Atoms and ions in minerals tend to become packed together as closely as their sizes permit. In an ionically bonded mineral, each ion attracts as many oppositely charged ions as can fit around it; their relative sizes therefore determine how many negative ions will surround a positive ion, and vice versa. In the mineral halite (NaCl), for example, one relatively small sodium ion (Na^+) is always surrounded by six larger chlorine ions (Cl^-) (Fig. 2-10).

Ionic Substitution Certain ions of similar size and charge can replace one another within a crystal structure, depending on which is most readily available during the mineral's formation. As a result of such *ionic substitution*, some minerals that have the same internal arrangement of ions may

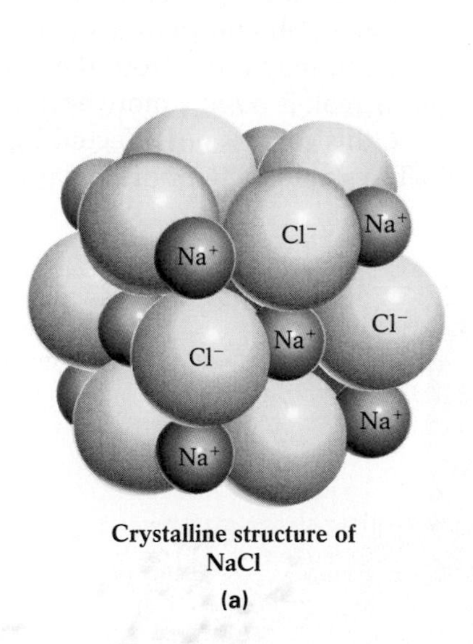

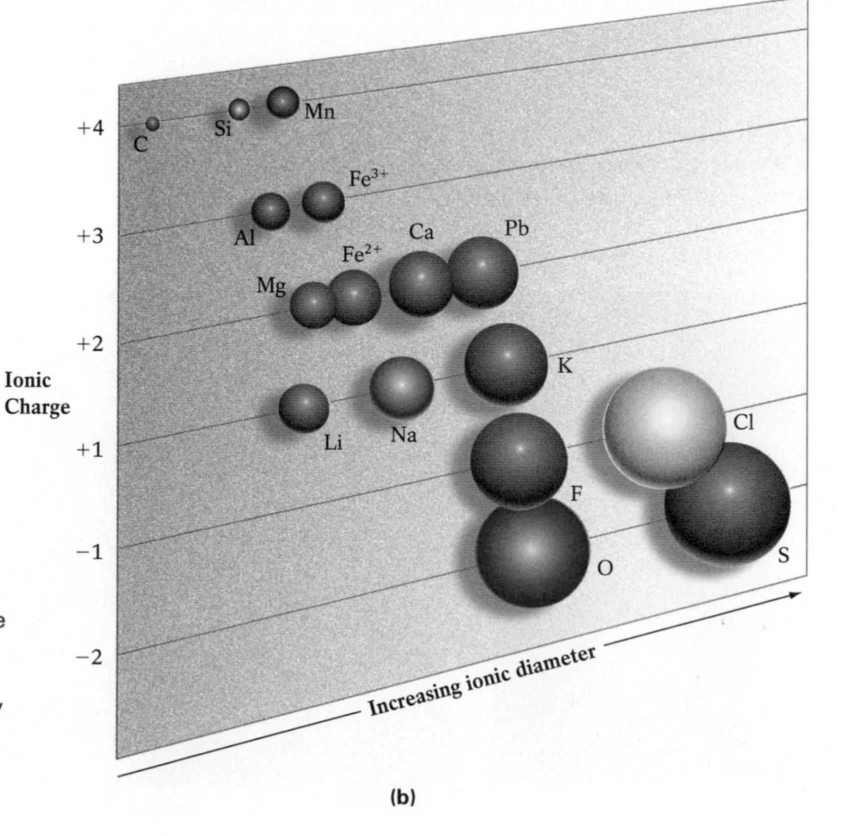

Figure 2-10 Mineral formation. **(a)** The crystalline structure of halite (NaCl), in which small sodium ions (Na^+) are tucked in between larger chlorine ions (Cl^-). **(b)** Positive ions are generally smaller than negative ions, because they tend to lose their outermost energy level in bonding; in a crystal structure, therefore, smaller positive ions usually occupy the spaces between larger negative ions.

exhibit minor variations in composition. For example, iron (Fe^{2+}) and magnesium (Mg^{2+}), which are nearly identical in size and charge, substitute freely for one another in the mineral olivine ($(Fe,Mg)_2SiO_4$). The color, melting point, and other physical characteristics of olivine differ depending on whether Fe^{2+} or Mg^{2+} is predominant, though olivine's chemical stability and crystal structure remain unaffected.

Polymorphism Two minerals may have the same chemical composition but different crystal structures because they formed under different temperature or pressure conditions. Such minerals are known as **polymorphs** ("many forms"). Graphite (the "lead" in your pencil) and diamond, for example, are polymorphs that consist entirely of carbon. Graphite forms under the low pressure prevalent at shallow depths (only a few kilometers below the Earth's surface), whereas diamond's much more compressed structure results from its formation under intense pressure at depths greater than 150 kilometers (90 miles).

Identification of Minerals

A mineral's chemical composition and crystal structure give it a unique combination of chemical and physical properties that geologists can use to distinguish it from all other minerals. A mineral can seldom be identified accurately on the basis of only one of these properties; usually several must be established before a conclusive identification is made.

In the Field

Many of a mineral's properties are instantly apparent or can be ascertained with minimal effort or rudimentary technology. Most geologists and dedicated rockhounds can identify a great many minerals in the field by merely examining them with the naked eye and performing some very simple tests.

Color Although color may be the first thing you notice about a mineral, it is perhaps the least reliable identifying characteristic. Many minerals occur in a variety of colors due to impurities in their crystal structures (Fig. 2-11). Other minerals with similar colors have completely different compositions. For example, specimens of quartz, calcite, fluorite, halite, and gypsum may appear nearly identical. Because a mineral's color is rarely unique to that mineral, and because it can vary greatly, color is not by itself a reliable criterion for mineral identification.

Luster Luster describes how a mineral's surface reflects light. Minerals can exhibit metallic or nonmetallic luster. Minerals with a metallic luster are either shiny (like car bumpers or aluminum foil) or have the appearance of an oxidized metal (for example, limonite and hematite often look like rusted lumps of metal) (Fig. 2-12). When any light shines on a nonoxidized metal, the light energy stimulates the metal's loosely held electrons and causes them to vibrate. The vibrating electrons emit a diffuse light, giving the metallic surface its characteristic shiny luster. Minerals with nonmetallic lusters are more varied; they can be vitreous (glassy), pearly, silky, adamantine ("like a diamond"), dull, or earthy (Fig. 2-13).

Streak Streak is the color of a mineral in its powdered form, obtained when the mineral's surface is pulverized by rubbing it across an unglazed porcelain slab known as a streak plate. The color of a mineral's streak may differ from the color of the intact mineral sample. Streak is often a more accurate indicator of identity, because this color is not affected by trace impurities in the sample. The steel-gray—or often dark red—

Figure 2-11 Two mineral samples of quartz, which is composed of silicon dioxide (SiO_2). Their colors differ because they each contain minute traces of different impurities.

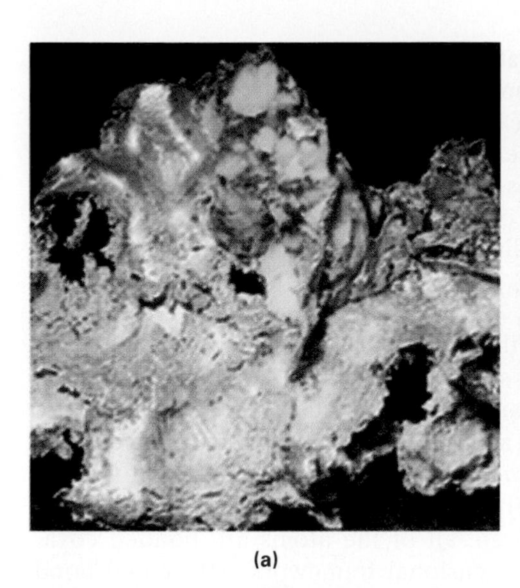

(a)

(b)

Figure 2-12 Two types of metallic luster. **(a)** Shiny, as in gold. **(b)** Nonshiny, as in limonite.

mineral hematite (Fe_2O_3), for instance, always has a distinctive reddish-brown streak (Fig. 2-14).

Hardness Geologists define hardness as a mineral's resistance to scratching or abrasion. To test hardness, the unknown mineral is scratched with a series of minerals or other substances of known hardness. (Geological hardness is *not* a function of how easily a mineral breaks—a solid rap with a hammer will easily shatter a diamond, the world's hardest natural substance.) Because every scratch mark represents the removal of atoms from the surface of the mineral, and thus the breakage of the bonds holding these atoms, a mineral's hardness indicates the relative strength of its bonds. Graphite, whose layers of covalently bonded carbon atoms are only weakly attached to each other by van der Waals bonds, is one of the softest minerals.

The Mohs Hardness Scale, named for its developer, German mineralogist Friedrich Mohs (1773–1839), assigns relative hardnesses to several common and a few rare and precious minerals. An unknown mineral that can be scratched by topaz but not quartz has a hardness between 7 and 8 on

(a)

(b)

Figure 2-13 Two types of nonmetallic luster. **(a)** Vitreous (glassy), as in rose quartz. **(b)** Pearly, as in feldspar.

Figure 2-14 Using streak to identify hematite. Though samples of hematite (Fe_2O_3) are usually steel-gray, hematite's streak is always reddish brown.

Table 2-2 The Mohs Hardness Scale

Mineral	Hardness	Hardness of Some Common Objects
Talc	1	
Gypsum	2	
		Human fingernail (2.5)
Calcite	3	
		Copper penny (3.5)
Fluorite	4	
Apatite	5	
		Glass (5–6), Pocketknife blade (5–6)
Orthoclase (potassium feldspar)	6	
		Steel file (6.5)
Quartz	7	
Topaz	8	
Corundum	9	
Diamond	10	

the Mohs scale. Table 2-2, which lists the minerals and common testing standards used in the Mohs scale, explains why geologists are often found with a few copper pennies, a pocketknife, and well-worn fingernails.

Cleavage Cleavage is the tendency of some minerals, when hammered or struck, to break consistently along distinct planes in their crystal structures where their bonds are weakest. The resulting mineral fragments possess smooth, flat surfaces, with consistent angles between adjacent surfaces. When a mineral that tends to cleave is struck along a plane of cleavage, every fragment that breaks off will have the same general shape. For example, shattering a cube of halite (table salt) produces numerous smaller cubes of halite, all with six sides at right angles to each other. Geologists often use the distinctive number of cleavage surfaces, and particularly the angles by which adjacent surfaces are joined, in identifying minerals (Fig. 2-15).

Fracture Minerals that do not cleave—because all of their bonds are equally strong—will break at random, or *fracture.* Unlike a straight, smooth-faced cleavage surface, a fracture appears as a jagged irregular surface or as a curved, shell-shaped (*conchoidal;* pronounced "kon-KOID-al") surface. In the mineral quartz, which is composed exclusively of silicon and oxygen, all of the atoms are bonded covalently in a three-dimensional framework with equal bond strengths in all directions. Thus, when a crystal of quartz is struck, it fractures (Fig. 2-16). Knowing that quartz fractures instead of cleaves enables geologists to distinguish it from similar-looking minerals, such as halite or calcite, that cleave.

Smell and Taste Experienced geologists occasionally sniff and lick rocks to help identify minerals that have a distinctive smell or taste. Some sulfur-containing minerals, for example, emit the familiar rotten-egg stench associated with hydrogen sulfide gas (H_2S). Halite's salty taste distinguishes it from similar-looking minerals such as quartz and calcite; sylvite (KCl) is distinctively bitter. Kaolinite absorbs liquid rapidly—when licked, it absorbs saliva and sticks to the tongue. Novices should be wary of tasting unknown minerals, however, as some can be dangerous. Realgar and orpiment, for example, which smell like garlic, especially when

(a)

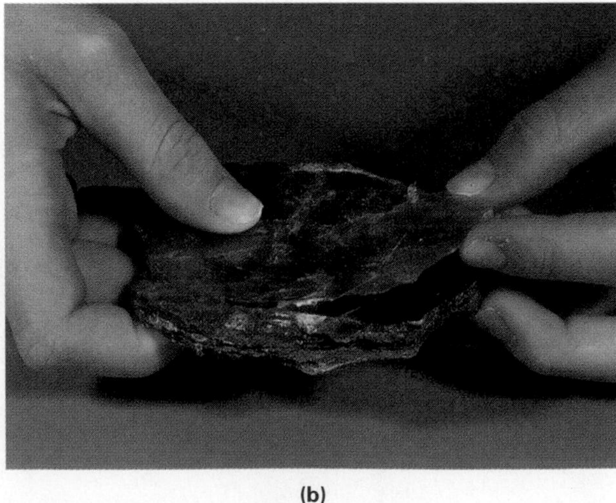

(b)

Figure 2-15 Distinguishing minerals by their cleavage. (a) Halite has three mutually perpendicular cleavage planes, forming cubes. **(b)** Mica has one perfect cleavage plane, forming sheets.

Figure 2-16 A conchoidal fracture surface on a quartz crystal. Quartz, with equally strong covalent bonds in all directions, has no planes of weakness. It therefore fractures irregularly instead of cleaving.

heated, have the poisonous metal arsenic as their major element.

Effervescence Certain minerals—particularly those that contain carbonate ions (CO_3^{2-})—effervesce, or fizz, when mixed with an acid. A few drops of dilute hydrochloric acid (HCl) on calcite ($CaCO_3$) produce a rapid chemical reaction that releases carbon dioxide gas (in the form of bubbles) and water; this property helps distinguish calcite from similar-looking minerals, such as quartz and halite, that do not effervesce.

Crystal Form Although minerals usually do not grow into perfect crystals, crystal form can nevertheless be an important factor in identifying a mineral. Its role is especially critical in the case of some minerals that grow into particularly distinctive forms, such as rosette-shaped barite and needle-shaped stibnite (Fig. 2-17).

In the Laboratory

Although geologists can positively identify many minerals in the field, some minerals have such similar physical characteristics that even experienced geologists will initially make only an educated guess as to their identity and then bring samples back from the field to test them in the laboratory. There, special equipment is available to analyze a variety of physical and chemical properties with greater precision than is ever possible in the field.

Specific Gravity Specific gravity is the ratio of a substance's weight to the weight of an equal volume of pure water. For example, a mineral that weighs four times as much as an equal volume of water has a specific gravity of 4. The precise specific gravity of an unknown mineral is generally determined in a laboratory, although relative specific gravity can also be helpful in the field in distinguishing between two apparently similar minerals. Minerals with a markedly higher specific gravity will feel much heavier relative to their size than minerals with a lower specific gravity will. For example, gold, with a specific gravity of 19.3, feels much heavier than "fool's gold" (the mineral pyrite), which has a specific gravity of only 5.

Other Laboratory Tests When exposed to ultraviolet light, certain minerals glow in distinctive colors. This property, called *fluorescence,* characterizes fluorite, calcite, scheelite ($CaWO_4$), and willemite (Zn_2SiO_4), among other minerals

(a)

(b)

Figure 2-17 Unusual crystal aggregates. (a) Stibnite needles. **(b)** Barite rosettes.

(Fig. 2-18). A mineral that continues to glow after removal of the ultraviolet light exhibits *phosphorescence.*

An *electron probe* is used to rapidly analyze extremely small mineral samples. This instrument beams electrons at the sample and then analyzes the distinctive X-rays emitted by the sample.

In *X-ray diffraction,* X-rays passed through a mineral sample become scattered, or *diffracted,* producing distinctive patterns (diffractograms) on an X-ray film. These patterns are determined by the arrangement of the atoms and ions in a mineral's crystal structure, and so are unique to each mineral.

Some Common Rock-Forming Minerals

Of the 92 naturally occurring elements in the Earth's continental crust, the eight most abundant (see Table 2-1) form the vast majority of minerals that compose the Earth's rocks. Most such **rock-forming minerals** are classified into one of six categories based on their chemical composition: the silicates, carbonates, oxides, sulfides, sulfates, and native elements.

(a)

(b)

Figure 2-18 Fluorescence. The fluorescent minerals willemite and calcite in this rock specimen **(a)** glow bright green and red, respectively, while exposed to ultraviolet light **(b)**.

Silicates

Most minerals in the Earth's crust and mantle are oxygen- and silicon-based compounds called **silicates.** Because silicon and oxygen are so abundant and unite so readily, the silicates—encompassing more than 1000 different minerals—make up more than 90% of the mass of the crust. They are the dominant component of most rocks, whether igneous, sedimentary, or metamorphic.

The crystal structure of all silicates contains repeated groupings of four negatively charged oxygen ions congregated around a single positively charged silicon ion to form a four-faced structure called a **silicon-oxygen tetrahedron** (SiO_4). As Figure 2-19 shows, the four oxygen ions in a silicon-oxygen tetrahedron each have a -2 charge, and the one silicon ion has a $+4$ charge. As a result, the tetrahedron carries an overall charge of -4. To form an electrically neutral compound, a silicon-oxygen tetrahedron must either acquire four positive charges by bonding with positive ions or disperse its negative charge by sharing its oxygen ions with neighboring tetrahedra.

The crystal structures of all silicate minerals are derived from one of five principal arrangements of the silicon-oxygen tetrahedra: independent tetrahedra, single chains, double chains, and three-dimensional framework structures. Each of these arrangements represents a different means of sharing oxygen ions.

Independent tetrahedra bond with positive ions of other elements to neutralize their charge. They share no oxygen ions, and so have a silicon-to-oxygen ratio of 1:4 (Fig. 2-20a). The most prominent mineral of this type is olivine,

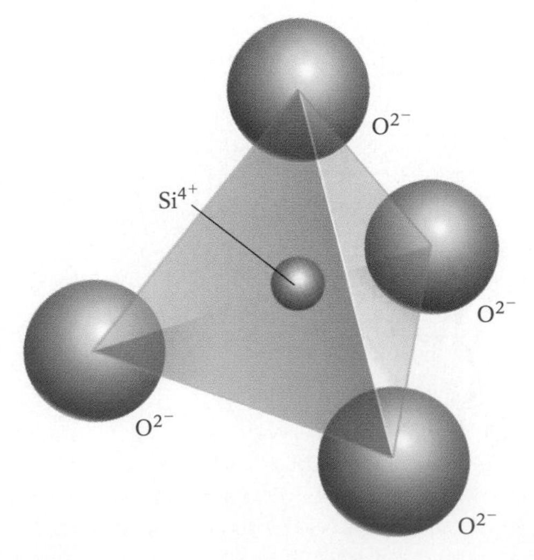

Figure 2-19 The silicon-oxygen tetrahedron. Four oxygen ions occupy the corners of this structure, with a lone silicon ion embedded in the open space at the center.

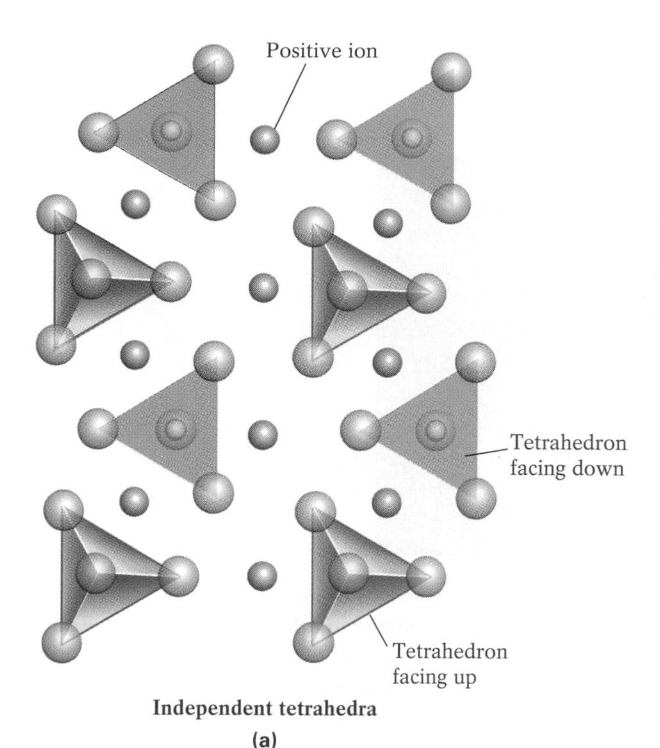

Figure 2-20 (a) Independent tetrahedra. Positive ions are positioned between tetrahedra such that each tetrahedron, with a −4 charge, bonds to two positive ions, each with a +2 charge, thereby neutralizing their combined charge. No oxygen ions are shared between the tetrahedra; therefore the silicon-to-oxygen ratio is 1:4. Photo: Olivine. As is typical of silicates with independent tetrahedral structures, this mineral fractures rather than cleaving.

in which iron (Fe^{2+}) and/or magnesium (Mg^{2+}) ions balance the negative charge of the tetrahedra.

Single chains of tetrahedra form when each tetrahedron shares two corner oxygen ions, producing a silicon-to-oxygen ratio of 1:3 (Fig. 2-20b). The most prominent group of minerals of this type encompasses the pyroxenes, which typically contain iron and/or magnesium ions that bind the chains together.

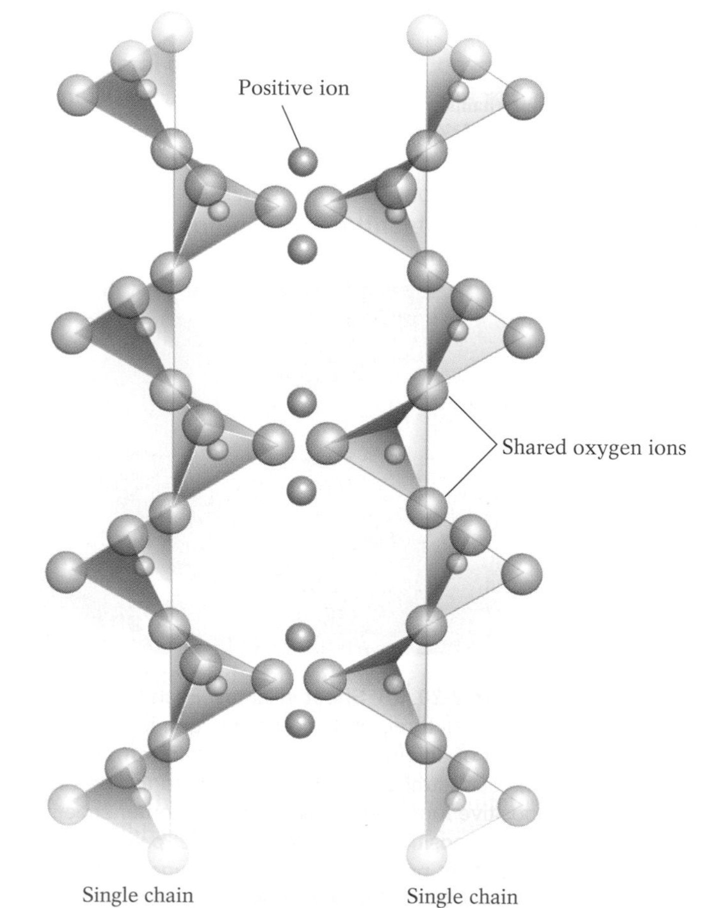

Figure 2-20 (b) Single chains. Each tetrahedron shares two of its corner oxygen ions with adjacent tetrahedra, forming a linear chain of tetrahedra with a silicon-to-oxygen ratio of 1:3. Because each tetrahedron still has a −2 charge after sharing two of its oxygen ions, the cumulative negative charge on such chains attracts a variety of positive ions that bind between them, neutralizing the negative charge of the chains and joining them loosely together. Photo: Pyroxene, showing the 90° cleavage angles characteristic of single-chain silicates.

Highlight 2-1 Synthetic Gems: Can We Imitate Nature?

Because the most valuable gemstones are rare, people have tried for centuries to duplicate nature's feat and produce synthetic gems. In the twentieth century, they have had some success, even managing to surpass nature in some cases. Artificial emerald crystals, first created in the 1930s, are more transparent, richer in color, and more perfect in shape than natural ones, which are often marred by gas bubbles and other impurities. The high quality of synthetic emeralds contributes to their great market value, which, at several hundred dollars per carat, is still far less than the price of the extremely rare natural ones.

Even diamonds can now be made in the laboratory, by subjecting carbon to extreme heat and pressure. (Almost any carbon-rich substance will do as a starting point—even sugar or peanuts.) On December 12, 1954, scientists in the General Electric research lab in Schenectady, New York, created the first tiny synthetic diamonds by subjecting carbon to great pressure and temperatures exceeding 3000°C (5400°F). Today, more than 20 tons of industrial-grade synthetic diamonds are produced each year, destined for such practical uses as drill bits in oil-well drilling and modern dentistry (Fig. 1).

In 1970, the first gem-quality diamonds were synthesized. Now, some synthetics even outshine the original. Strontium titanate, a synthetic mineral sold under the trade names of Fabulite and Wellington Diamond, glitters four times more vividly than a real diamond. Being only moderately hard, between 5 and 6 on the Mohs scale, it is not very durable, however. And because its creators can manufacture tons of it, this synthetic gem is only as rare as they choose. Another synthetic, cubic zirconia (Fig. 2), can be manufactured in batches of 50 kilograms (110 pounds) and sold wholesale for a few cents per carat. Its optical qualities are virtually indistinguishable from those of nature's diamonds, and cubic zirconia is quite durable. Only the fact that its specific gravity is higher than that of natural diamond reveals its identity.

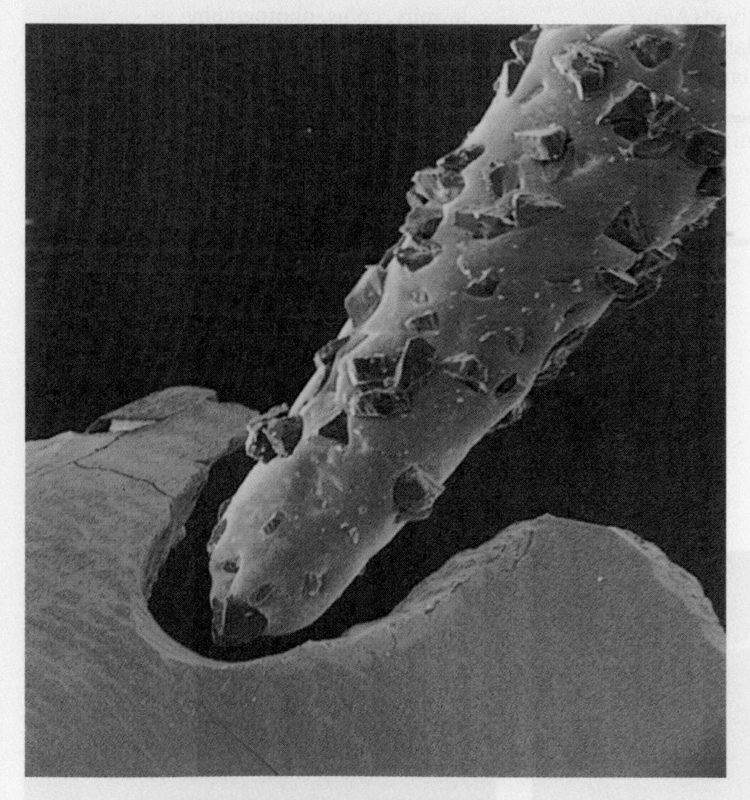

Figure 1 Industrial-grade synthetic gems. A drill bit studded with synthetic diamonds cuts swiftly through the relatively soft enamel of a human molar composed largely of the calcium-phosphorus mineral, apatite. (Magnified 25×)

Figure 2 Cubic zirconia, a synthetic diamond. Cubic zirconia may look exactly like a real diamond (compare with Fig. 2-22), but it is manufactured in bulk and quite cheaply.

Figure 2-22 Diamond before and after cutting. **(a)** A raw, uncut diamond. **(b)** A cut and faceted diamond. Actually, diamonds are generally cleaved (not "cut") along the planes of their weakest bonds. Knowing the sites of these weaker bonds, an experienced diamond cutter can cleave a dull, irregular-shaped diamond into a glittering, perfectly symmetrical jewel.

Molten rock often migrates into fractures in surrounding cooler rocks, where its ions and atoms crystallize in reasonably large spaces, producing perfect crystals and, if the space is large enough, enormous crystals. For example, a single pyroxene crystal excavated in South Dakota was more than 12 meters (40 feet) long, 2 meters (6.5 feet) wide, and weighed more than 8000 kilograms (8 tons). The same crystallization process also produces such complex silicate gemstones as topaz, tourmaline, and beryl.

Alternatively, gemstones may form when the heat and pressure applied along the edges of colliding tectonic plates cause the ions and atoms in their rocks to migrate and recombine, creating new minerals that are more stable under the new conditions. For example, heating and compression of carbonate rocks that contain aluminum ions can cause the aluminum to combine with oxygen liberated from calcite, forming the aluminum oxides we know as rubies and sapphires.

Diamonds are transformed from carbon under extreme pressure. These gems are most often found where superheated gas has propelled molten rock rapidly from depths greater than 150 kilometers (90 miles) to the surface, carrying the diamonds along with it. The resulting diamond-rich structures, known as kimberlite pipes (named for Kimberley, South Africa), are typically a few hundred meters to a kilometer across and many kilometers deep. Kimberlite pipes are found in Siberia, India, Australia, Brazil, the Northwest Territories of Canada, southern and central Africa, and the Rockies of Colorado and Wyoming. Most, however, do not yield gem-quality diamonds.

When gem-quality diamonds are found, they exist as raw diamonds. The ability to cut raw diamonds into the valuable multifaceted gemstones with which we are more familiar is a technically difficult and exacting skill. It requires steady hands, finely honed tools, and a precise knowledge of diamonds' crystalline structure (Fig. 2-22). Diamonds are also sawed (a slower but safer way to cut a stone) using ultrafine diamond-edged blades rotating at high speeds.

Now that we have introduced the basic structure of minerals, the methods by which they form, and techniques for identifying them, we can begin to examine more closely the common types of rocks that make up the Earth's crust. In the next five chapters, we discuss the formation of igneous, sedimentary, and metamorphic rocks (introduced in Chapter 1 with the rock cycle) and describe how the minerals in these rocks can be used to interpret past geologic events.

Chapter Summary

Minerals are naturally occurring inorganic solids with specific chemical compositions and specific internal structures. **Rocks** are naturally occurring aggregates of one or more minerals. Minerals are composed of one or more chemical **elements,** the form of matter that cannot be broken down to a simpler form by heating, cooling, or reacting with other elements. Each element, in turn, consists of **atoms,** infinitesimally small particles that retain all of an element's distinguishing chemical characteristics. When atoms of two or more elements combine in specific proportions, they form chemical **compounds,** which have properties that differ from the properties of any of their constituent elements individually. All minerals are chemical compounds.

In the center of an atom is its **nucleus,** which contains both positively charged particles called **protons** and uncharged particles of equal mass called **neutrons.** An element's

atomic mass is the sum of the masses of its protons and neutrons. An element's **atomic number** is determined by the number of protons in its nucleus; every atom of a given element has the same number of protons and thus the same atomic number. Atoms of a given element may differ in atomic mass, however, because the number of neutrons in their nuclei may vary. Atoms of a given element that contain different numbers of neutrons are called **isotopes** of that element.

An atom's nucleus is surrounded by a cloud of negatively charged particles called **electrons** that move about the nucleus at high speed. Each electron occupies a specific region of space called an **energy level;** an atom may have one or more energy levels.

Atoms combine to form chemical compounds in a variety of ways known as **bonding.** Two key factors determine which atoms will unite with which other atoms to form compounds: Each atom tends to achieve chemical stability by having its outermost energy level filled with electrons, and an electrically neutral compound is more stable. To attain a full outermost energy level, an atom may donate or acquire electrons, thereby becoming an electrically charged particle called an **ion.** An atom that donates electrons becomes positively charged; an atom that acquires electrons becomes negatively charged. Oppositely charged ions attract one another, forming an **ionic bond.**

Sometimes atoms share the electrons in their respective outer energy levels, forming a **covalent bond.** In **metallic bonding,** electrons move continually among numerous closely packed nuclei. Intermolecular bonds such as **hydrogen bonds** and **van der Waals bonds** form from weak attractions between molecules or groups of atoms, caused by uneven distribution of their moving electrons.

As a mineral forms through chemical bonding, all of its ions or atoms occupy specific positions to create a **crystal structure,** a three-dimensional pattern repeated throughout the mineral. When mineral growth is not limited by space, a **crystal** may form with a regular geometric shape that reflects the mineral's internal crystal structure. When molten rock cools too quickly for its atoms to form an orderly arrangement, the resulting solids, called **mineraloids,** lack a crystal structure and thus are not true minerals.

The types of minerals that will form at a given time and place are determined by which elements are available to bond, the charges and sizes of their ions, and the temperature and pressure under which the minerals form. **Polymorphs** are minerals that have the same chemical composition but different crystal structures because they form under different temperature or pressure conditions.

Geologists identify minerals primarily by noting external characteristics, such as color, luster, streak, hardness, cleavage, and fracture. They also measure physical properties such as specific gravity.

The Earth's crust is primarily composed of only eight elements, which combine to produce the **rock-forming minerals.** The two most prominent elements, oxygen and silicon, combine readily to form the **silicon-oxygen tetrahedron.** It serves as the basic building block of the Earth's most abundant group of minerals, the **silicates,** which make up more than 90% of the mass of the Earth's crust. Silicon-oxygen tetrahedra may be linked in a variety of crystal structures: independent tetrahedra (such as in olivine), single chains of tetrahedra (such as in pyroxene), double chains (such as in amphibole), sheet structures (such as in mica), and framework structures (such as in feldspar and quartz).

A number of nonsilicates are also common rock-forming minerals. They include the **carbonates,** the **oxides,** the **sulfides** and **sulfates,** and the **native elements.**

Gemstones are minerals that are valued for their particularly appealing color, luster, or crystal form. Some gemstones, such as diamonds and emeralds, are quite rare; others are unusually well-formed specimens of relatively commonplace minerals, usually containing trace impurities in their structures that impart distinctive color to the crystals.

Key Terms

minerals (p. 20)
rocks (p. 20)
element (p. 20)
atoms (p. 20)
compounds (p. 20)
protons (p. 20)
neutrons (p. 20)
nucleus (p. 20)
electrons (p. 20)
atomic mass (p. 21)
atomic number (p. 22)
isotopes (p. 22)
energy level (p. 22)
bond (p. 22)
ion (p. 23)
ionic bonds (p. 23)
covalent bond (p. 23)
metallic bonding (p. 24)
hydrogen bonds (p. 24)
van der Waals bond (p. 24)
crystal (p. 24)
crystal structure (p. 24)
mineraloids (p. 24)
polymorphs (p. 26)
rock-forming minerals (p. 30)
silicates (p. 30)
silicon-oxygen tetrahedron (p. 30)
carbonates (p. 33)
oxides (p. 34)
sulfides (p. 34)
sulfate (p. 34)
native elements (p. 34)

Questions for Review

1. What is a mineral? How does a mineral differ from a rock?

2. Briefly describe the structure of an atom. What is an isotope? What is an ion?

3. How does an atom achieve chemical stability? How does a chemical compound achieve electrical neutrality?

4. Describe three types of chemical bonding.

5. What is a mineral crystal? Describe two circumstances under which a mineral probably will not grow into a well-formed crystal.

6. Define and give an example of mineral polymorphs.

7. Describe four properties of minerals that could help you to identify an unknown mineral.

8. Briefly discuss why the silicate minerals are the most abundant in nature.

9. List four different silicate structures and give a specific mineral example of each.

10. List two types of common nonsilicates and give a specific mineral example of each.

11. Describe the connections between at least two different geological environments and the formation of gemstones.

For Further Thought

1. If some sodium (Na^+) substitutes for calcium (Ca^{2+}) in the plagioclase feldspars, why must some aluminum (Al^{3+}) replace some silicon (Si^{4+}) in the mineral's structure?

2. Sulfur forms a small ion with a high positive charge. Why doesn't sulfur unite universally with oxygen to form the basic building blocks of most crustal minerals?

3. Why doesn't the mineral quartz exhibit the diagnostic property of cleavage? Considering its physical beauty, why isn't a quartz crystal a more valuable gemstone?

4. The photos at the right show crystals of real gold and "fool's gold" (pyrite). How would you distinguish between these two similar-looking minerals?

3

Igneous Processes and Igneous Rocks

On the island of Hawai'i, you can watch a volcano erupt and later touch the warm rock that passed through the volcano in a molten state just hours or days earlier. In the Sierra Nevada mountains of California, you can walk on rocks that cooled 80 million years ago from molten rock 20 kilometers (12 miles) below the Earth's surface. As we saw in Chapter 1, such rocks, which cooled and crystallized directly from molten rock, either at the surface or deep underground, are called **igneous rocks** (from the Latin *ignis,* meaning "fire"). More than 95% of the Earth's outer 50 kilometers (30 miles) consists of igneous rocks. In fact, the remains of ancient volcanic eruptions and vast uplifted regions of formerly subsurface igneous rocks can be found in almost every state and province of North America.

Geologists can observe molten material spewing from a volcano, but they cannot see it moving underground—and almost all igneous rock solidifies far beneath the surface. One way that geologists investigate underground igneous processes is to look for regions where erosion has removed surface rock layers (that is, by physically wearing away rocks by natural agents such as wind, water, and ice), thus opening up windows through which we can see the subsurface. Igneous rocks that solidified underground appear in some of North America's most scenic places, from Mount Katahdin in northern Maine to the Yosemite Valley of eastern California (Fig. 3-1). Some of our continent's most ancient rocks are the igneous roots of one-time mountains that have long since eroded to expose the underlying rocks. Such rocks can be seen in northern Minnesota, Ontario, and Quebec.

Some geologists investigate igneous processes by simulating them in laboratories. This practice allows geologists to observe the effects of pressure, temperature, composition, and other factors on the melting and crystallization points of sample rocks and minerals.

Our discussion of the Earth's igneous processes and rocks first describes some of the characteristics of molten rock, before turning to the types of rocks that form when molten material cools and solidifies. We also examine how and why rocks melt and crystallize, paying special attention to how plate tectonics affects the origin and distribution of igneous rocks on Earth (and how Moon rocks differ in this

Figure 3-1 Igneous rocks. These rocks exposed along the western end of Lake Superior are part of the Duluth lopolith, a huge expanse of igneous rock that solidified underground millions of years ago.

respect). Finally, we investigate the economically valuable materials—such as gold, silver, and copper—that originate from igneous processes.

What Is Magma?

Magma is molten rock that flows within the Earth. It may be completely liquid or, more commonly, a fluid mixture of liquid, solid crystals, and dissolved gases. When magma reaches the Earth's surface, we call it **lava,** or molten rock that flows above ground.

Magma forms when underground temperatures become high enough to break the bonds in some minerals, causing the minerals to melt. The rock then changes from a crystalline solid to a fluid mix containing freely moving ions and atoms as well as some still-solid crystalline fragments. Different minerals melt out of the rock at different temperatures as the heat gradually increases, with the minerals having the highest melting points remaining the longest as still-solid fragments in the magma. At the same time, the composition of the magma changes as each newly molten mineral enters and enriches it.

When heat dissipates from a magma, its bonds no longer break and new bonds start to form. First, some of the free atoms and ions in the liquid bond to form tiny crystals. Additional ions and atoms bond at prescribed sites in the crystal structures. The crystals grow until they touch the edges of adjacent crystals. As cooling progresses, different minerals crystallize from the magma, again changing the magma's composition. If cooling continues long enough, the entire body of magma will become solidified as igneous rock.

Classification of Igneous Rocks

Igneous rocks are classified based on their two most obvious properties: their texture, which is determined by the size and shape of their mineral crystals and the manner in which these grew together during cooling, and their composition, which is determined by the minerals that they contain.

Figure 3-2 Phaneritic rocks. Rocks that solidify slowly underground, as did this granite in Yosemite National Park, California, have phaneritic (coarse-grained) textures.

Igneous Textures

A rock's *texture* refers to the appearance of its surface—specifically the size, shape, and arrangement of the rock's mineral components. The most important factor controlling these features in igneous rocks is the rate at which a magma or lava cools. When a magma's minerals crystallize slowly underground over thousands of years, crystals have ample time to grow large enough to be seen clearly with the unaided eye. The resulting texture is called *phaneritic* (pronounced "FAN-er-i-tic"; from the Greek *phaneros,* meaning "visible") (Fig. 3-2). Slow cooling occurs when magmas enter, or *intrude,* preexisting solid rocks; thus rocks with phaneritic textures are known as **intrusive rocks.** They are also called **plutonic rocks** (pronounced "ploo-TON-ic"; for Pluto, the Greek god of the underworld).

Some igneous rocks develop at relatively low temperatures from ion-rich magmas containing a high proportion of water. Under these conditions, ions move quite readily to bond with growing crystals, enabling the crystals to become unusually large (sometimes several meters long). Rocks with such exceptionally large crystals are called *pegmatites* (from the Greek *pegma,* meaning "fastened together") (Fig. 3-3). In western Maine, near the towns of Bethel and Rumford, some pegmatitic rocks contain 5-meter (17-foot)-long crystals of the mineral beryl.

Some igneous rocks solidify from lava so quickly that their crystals have little time to grow. These *aphanitic* (pronounced "af-a-NIT-ic"; from the Greek *a phaneros,* meaning "not visible") rocks have crystals so small that they can barely be seen with the naked eye (Fig. 3-4). Rocks with aphanitic textures are called **extrusive rocks,** because they form from lava that has flowed out, or been *extruded,* onto the Earth's surface. They are also known as **volcanic rocks,** because lava is a product of volcanoes (named for Vulcan, the Roman god of fire).

In some igneous rocks, large, often perfect, crystals are surrounded by regions with much smaller or even invisible

Figure 3-3 Pegmatites. Extremely coarse-grained pegmatites, such as the one shown here, form from ion-rich magmas having a high water content.

Figure 3-4 Volcanic rocks. Volcanic rocks, such as this basalt, typically have aphanitic (very small-grained) textures, because they solidify rapidly above ground.

Figure 3-5 Porphyritic rocks. Some rocks have a porphyritic texture, marked by large crystals surrounded by an aphanitic matrix.

Figure 3-6 Glassy volcanic rocks. Obsidian **(a)** and pumice **(b)** contain no crystals because they solidify instantaneously. Pumice, which forms from lava foam, commonly has so many tiny air-filled cavities that it can float in water.

grains (Fig. 3-5). These *porphyritic* (pronounced "por-fa-RIT-ic") textures are believed to form as a result of slow cooling followed abruptly by rapid cooling. First, gradual underground cooling produces large crystals that grow slowly within a magma. Next, the mixture of remaining liquid magma and the early-formed crystals is forced close to the surface or actually escapes into the air. There the liquid cools rapidly to produce the body of smaller grains that envelops the larger crystals.

When lava from a volcano erupts into the air or flows into a body of water, much of it cools so quickly that its ions don't have time to become organized into any crystals. The texture of the resulting rock is described as *glassy*. Two common types of volcanic glass exist, both produced by instantaneous cooling of lava: dark-colored *obsidian* and light-colored, cavity-filled *pumice* (Fig. 3-6). The latter forms from bubbling, highly gaseous lava foam.

Igneous Compositions

The Earth's magmas consist largely of the most common elements: oxygen, silicon, aluminum, iron, calcium, magnesium, sodium, potassium, and sulfur. The relative proportions of these components found at any given time within a body of magma give the magma its distinctive characteristics and ultimately determine the mineral content of the rocks it will form. Igneous rocks and magmas are classified into four main compositional groups—ultramafic, mafic, intermediate, and felsic—based on the proportion of silica (oxygen and silicon) they contain (Table 3-1). Figure 3-7 shows how the mineral content of igneous rocks varies in these categories.

Ultramafic Igneous Rocks The term "mafic" is derived from *ma*gnesium and *f*errum (Latin for "iron"). Ultramafic igneous rocks are dark in color and very dense, because they are dominated by the iron- and magnesium-containing silicate

Table 3-1 Common Igneous Compositions

Composition Type	Percentage of Silica	Other Major Elements	Relative Viscosity of Magma	Temperature at Which First Crystals Solidify	Igneous Rocks Produced
Felsic	>65%	Al, K, Na	High	~600 – 800°C (1100–1475°F)	Granite (plutonic) Rhyolite (volcanic)
Intermediate	55–65%	Al, Ca, Na, Fe, Mg	Medium	~800–1000°C (1475–1830°F)	Diorite (plutonic) Andesite (volcanic)
Mafic	40–55%	Al, Ca, Fe, Mg	Low	~1000–1200°C (1830–2200°F)	Gabbro (plutonic) Basalt (volcanic)
Ultramafic	<40%	Mg, Fe, Al, Ca	Very low	>1200°C (2200°F)	Peridotite (plutonic) Komatiite (volcanic)

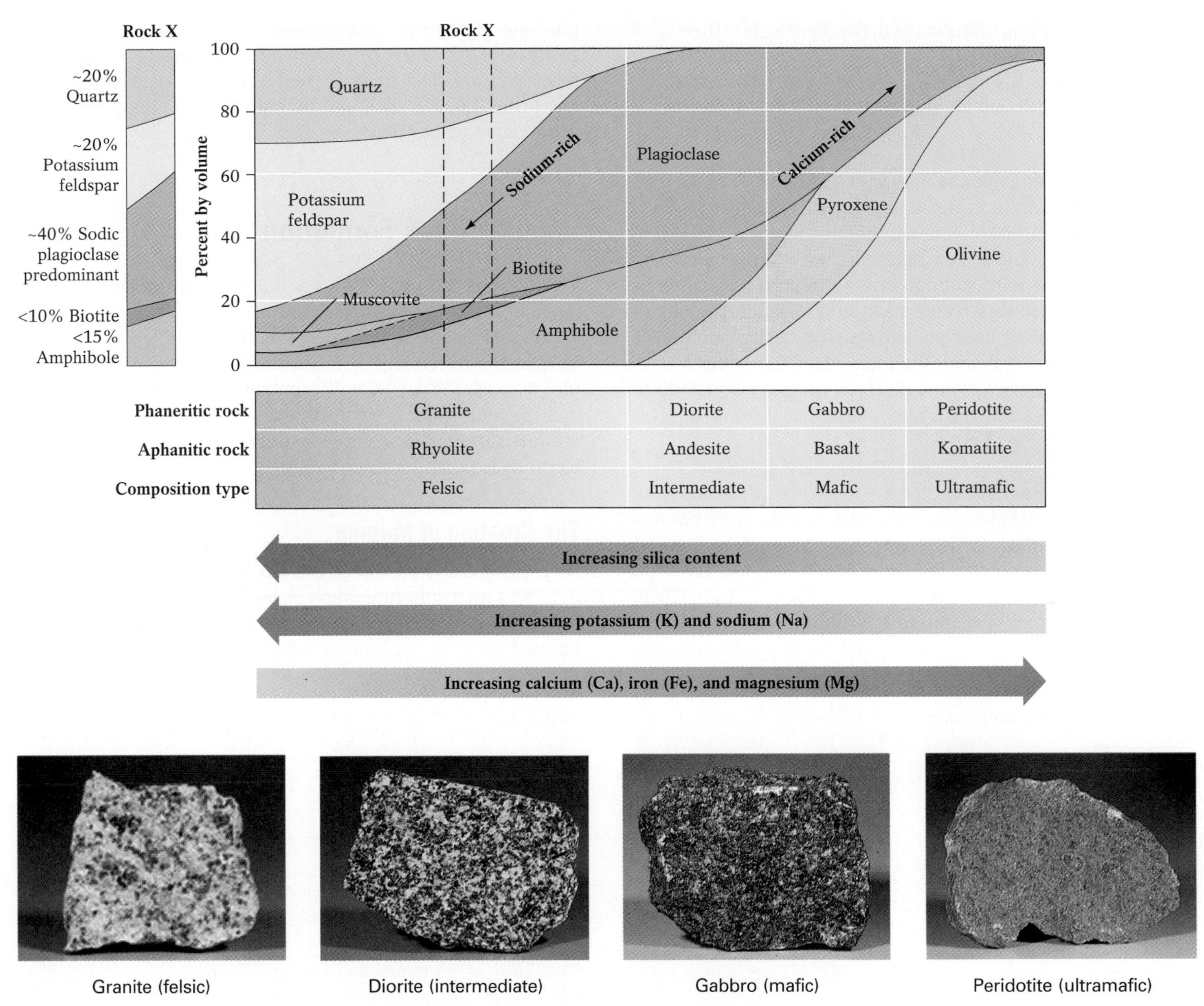

Figure 3-7 An igneous rock classification chart. Compositional types among the igneous rocks range from felsic to ultramafic. The mineral components of the rocks are indicated by colored areas in the body of the chart. (The sample segment shows how to interpret the chart, using as an example a rock falling between granite and diorite in composition.)

minerals (called *ferromagnesian* minerals) olivine and pyroxene and contain relatively little silica (less than 40%). The most common ultramafic rock, **peridotite** (pronounced "pe-RID-o-tite"), contains 40% to 100% olivine. Ultramafic rocks generally crystallize slowly deep within the Earth's interior and appear at the Earth's surface only where extensive erosion has removed overlying crustal rocks. They are most likely to be found near continental collision plate boundaries, where deep rocks have been uplifted.

Mafic Igneous Rocks Mafic igneous rocks have a silica content ranging between 40% and 55%, with the principal minerals being pyroxene, calcium feldspar, and a minor amount of olivine. They are the most abundant rocks of the Earth's crust; the aphanitic volcanic rock **basalt** (pronounced "ba-SALT") is the single most abundant of them. Most of the ocean floor and many islands, such as the Hawai'ian chain, are composed of basalt. Basalt also constitutes vast areas of our continents and is found in Brazil,

India, South Africa, Siberia, and the Pacific Northwest of North America (Fig. 3-8). The plutonic equivalent of basalt is **gabbro;** it has the same composition as basalt but, because it cools more slowly deep within the Earth, is coarse-grained.

Intermediate Igneous Rocks Intermediate igneous rocks contain more silica than mafic rocks, including between 55% and 65% silica, and are generally lighter in color. They typically consist of some ferromagnesians, such as pyroxene and amphibole, along with sodium- and aluminum-rich minerals such as sodium feldspar and mica, and a small amount of quartz. Examples of intermediate igneous rocks include the aphanitic volcanic rock **andesite** (named for the Andes Mountains of South America, where this type of igneous rock dominates the local geology) and its phaneritic plutonic equivalent **diorite** (see Fig. 3-8).

Felsic Igneous Rocks The term "felsic" is derived from *fel*dspar and *si*lica. Felsic igneous rocks contain more silica—65% or more—than either mafic or intermediate igneous rocks. They are generally lighter in color because they are poor in iron, magnesium, and calcium silicates, but rich in potassium feldspar, aluminum-rich (muscovite) mica, and quartz. Examples of felsic igneous rocks include the common plutonic phaneritic rock **granite** and its volcanic aphanitic equivalent, **rhyolite** (pronounced "RYE-uh-lite"). Rocks of felsic composition have a greater variety of textures than any other igneous rock and include several glassy rocks and ultracoarse pegmatites (see Fig. 3-8).

Igneous Rock Formation

The melting of solid rock in the Earth's interior to form magma and the recrystallization of the magma to form igneous rock is not as simple as the melting of ice to form water and the refreezing of water to form ice cubes. Ice is simple; it is composed of only one substance and has a single melting and freezing point. Interior rock is composed of numerous substances, and the melting and refreezing (crystallization) is much more complex.

The Creation of Magma

As rocks are heated within the Earth, not all minerals within the rocks melt simultaneously. A body of rock undergoes **partial melting** when it is heated to the melting points of some but not all of its component minerals. The magma produced

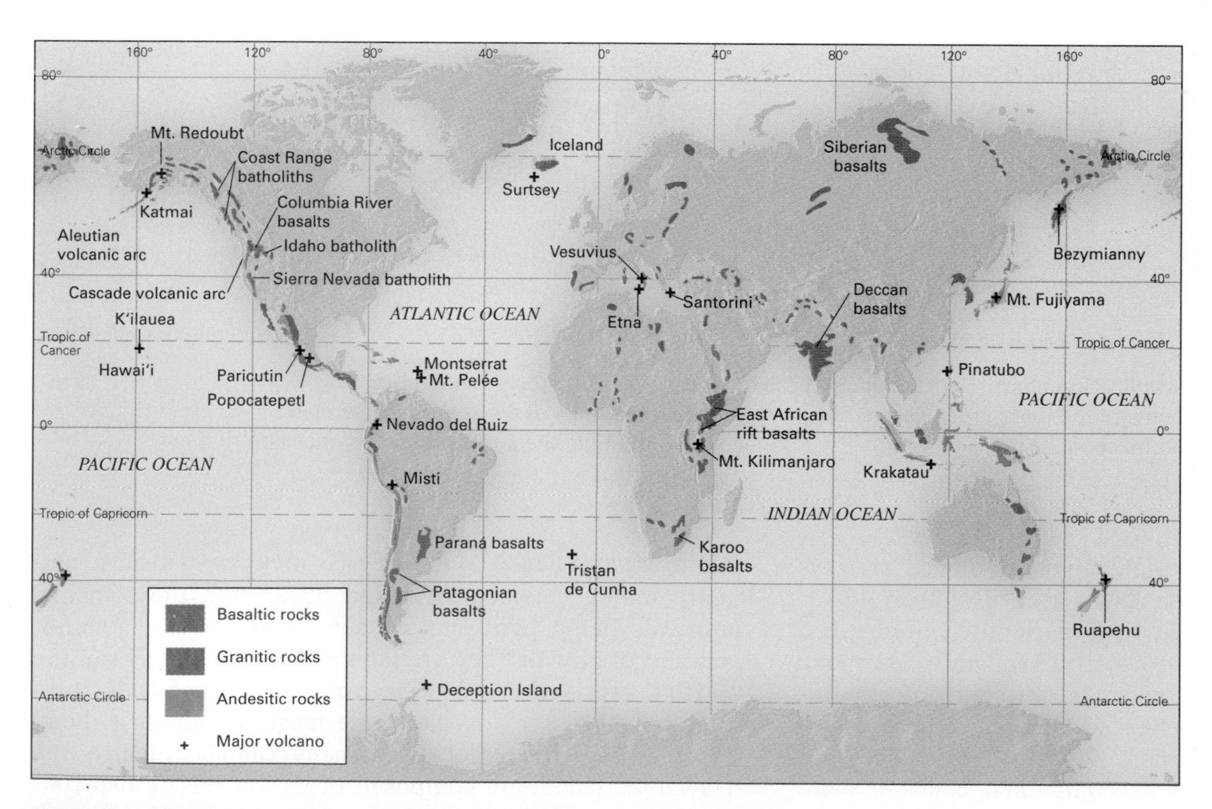

Figure 3-8 Distribution of the common terrestrial igneous rock types. As well as being the most abundant igneous rock of the ocean floors, the mafic igneous rock basalt is widespread on the continents.

by partial melting contains both molten minerals and still-solid chunks of other minerals with higher melting points. In particular, felsic minerals melt at a lower temperature than mafic minerals, so partial melting of mafic rocks may produce felsic or intermediate magmas. If temperatures are subsequently raised, a progressive melting of minerals with higher melting points will occur; these minerals will be added to the magma, thereby changing its composition.

If partial melting determines the composition of a magma, then several other factors determine where and when the magma will form. Heat, pressure, and the amount of water in rocks all combine to influence the point at which the rocks melt. As we will see, these factors affect the type of rocks that eventually form from magmas and their locations on Earth.

Heat The heat in the Earth's interior comes from three primary sources: the heat produced during the formation of the planet, which is still rising from the Earth's core; the heat liberated continuously by decay of radioactive isotopes; and the frictional heat produced as the Earth's plates move against one another and over the underlying asthenosphere. These heat sources ensure that temperatures in the Earth increase with depth, at a rate referred to as the *geothermal gradient.* The geothermal gradient is steepest from about 50 to 250 kilometers (30–150 miles) of depth. Here, temperatures exceed the 700°C (1300°F) required to melt most felsic minerals and the 1300°C (2400°F) required to melt most mafic minerals. Thus most magmas tend to form at these depths, particularly when the right combination of pressure and fluids accompanies the high temperatures.

Pressure Heat melts minerals by causing the ions and atoms in their crystal structures to vibrate, stretching and eventually breaking the bonds between them. When a crystal structure is under pressure, however, its ions remain in place longer. Higher temperatures are then required to break their bonds. Thus, the higher the pressure on a mineral, the higher its melting point. Because rocks located far beneath the Earth's surface experience great pressure from the weight of overlying rocks, higher temperatures are needed to melt them (Fig. 3-9). If the pressure on a rock somehow becomes reduced or removed—as happens when tectonic plates rift and diverge—its melting point drops below its current temperature and it begins to melt.

Water Water, even a small amount, weakens the bonds within minerals (see "Intermolecular Bonding," in Chapter 2) and thus lowers the melting point of rocks. Under high pressure, water has an even greater effect on the melting point of a mineral. Whereas dry rocks become more resistant to melting with greater depth, wet rocks become less resistant, because high pressure drives more water into the rocks. The combination of high pressure and high water content represents an important factor in the production of magmas at subducting plate boundaries.

The Crystallization of Magma

As magma cools and solidifies, it undergoes a number of significant changes in composition. Minerals that melt last (at the highest temperatures) during heating—the more mafic minerals—are the first to crystallize during cooling. A partially

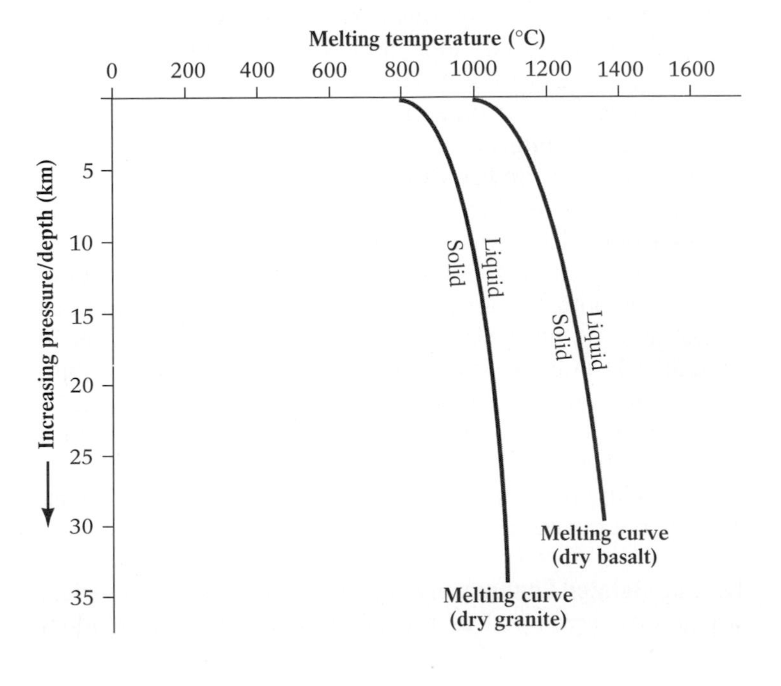

Figure 3-9 Melting-temperature curves for dry basalt and dry granite. (Adding water to rock changes its melting curve.) For both, melting temperatures increase with increasing depth, because the pressure at greater depths stabilizes rock's crystal structure, thereby raising its melting point.

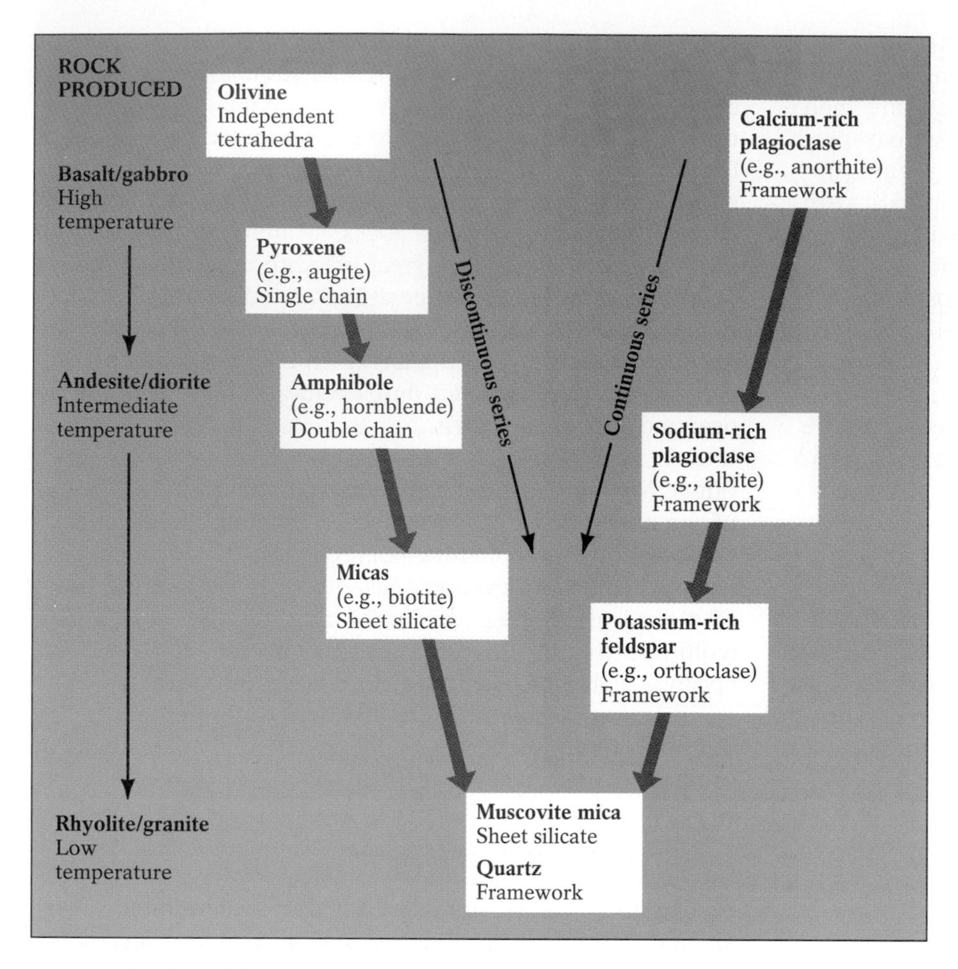

Figure 3-10 Bowen's reaction series. This series shows the sequence of minerals that crystallize as an initially mafic magma cools under conditions in which the early-forming crystals remain in contact with the still-liquid magma.

cooled magma may therefore contain solid crystals of mafic minerals along with liquid of a more felsic composition. As the magma continues to cool, additional ions and atoms crystallize, leaving behind progressively less liquid.

Bowen's Reaction Series In 1922, Canadian geochemist Norman Levi Bowen and his colleagues at the Geophysical Laboratory of the Carnegie Institution in Washington, D.C., determined the sequence in which silicate minerals crystallize as magma cools. Their work made it possible to summarize a complex set of geochemical relationships, called **Bowen's reaction series,** in a single diagram (Fig. 3-10), and demonstrated that a full range of igneous rocks, from mafic to felsic, could be produced from the same, originally mafic magma. The early-forming crystals remain in contact with the still-liquid parent magma, continue to react with it, and so evolve into new minerals.

Bowen's reaction series shows that the silicate minerals can crystallize from mafic magmas in two ways—in a discontinuous series or in a continuous series. Ferromagnesian minerals (the iron- and magnesium-rich silicates) crystallize one after another in a specific sequence. Because each successive type of ferromagnesian mineral crystallized differs in both composition and internal structure from the one before, Bowen called this progression the *discontinuous series.* As mafic magma cools, the first ferromagnesian mineral to crystallize is olivine, which has a low silica content and a relatively simple structure of independent tetrahedra. The crystallization of olivine removes iron and magnesium atoms and ions from the liquid portion of the magma, increasing the proportion of the other major ions in it. Meanwhile, the scattered olivine crystals growing in the magma continue to incorporate silica, and their tetrahedra become linked in the single-chain structure of the pyroxenes. The evolution of the ferromagnesian minerals continues as pyroxene crystals acquire more silica and become transformed into the double-chained amphiboles. Eventually, the series culminates in the formation of the complex sheet silicate biotite mica, the last ferromagnesian mineral to form. By then, the iron and magnesium ions and atoms in the liquid have been depleted. As a result, any minerals that crystallize after biotite will contain no iron or magnesium.

At the same high temperatures at which olivine and the pyroxenes crystallize, calcium feldspar crystallizes as well. As these early-forming crystals continue to interact with the remaining liquid, sodium ions from the liquid magma gradually replace the calcium ions in the calcium feldspar. Eventually, the growing crystals are completely converted to sodium feldspar. Because one type of ion is being replaced by a very similar ion, the internal structure of these minerals does not change; therefore Bowen called this progression the *continuous series.* The resulting sequence of plagioclase feldspars ranges from calcium-rich anorthite through a variety of intermediate calcium–sodium mixtures, to sodium-rich albite.

After both the ferromagnesian minerals (discontinuous series) and the plagioclase feldspars (continuous series) have crystallized completely from an initially mafic magma, less than 10% of the original liquid remains. Depending on its initial composition, this liquid may now contain high concentrations of silica, potassium, and aluminum. In such a case, potassium (alkali) feldspar, potassium-aluminum mica (typically muscovite), and quartz are the last minerals to crystallize.

Cooling-Related Changes in Magma Bowen's reaction series, which was developed in the laboratory, assumes an ideal

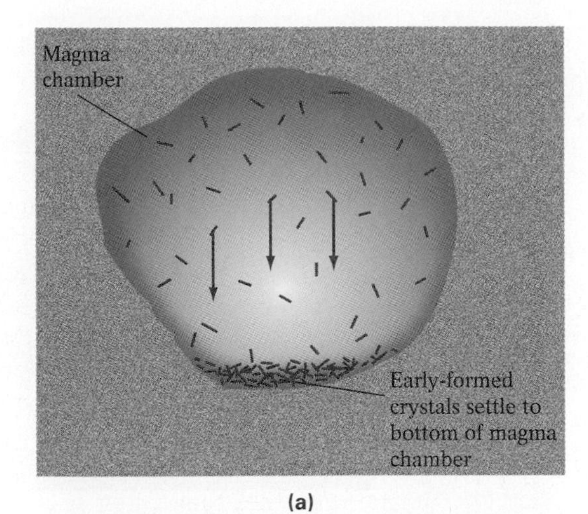

(a)

Crystals become affixed to magma chamber roof and walls as magma circulates

(b)

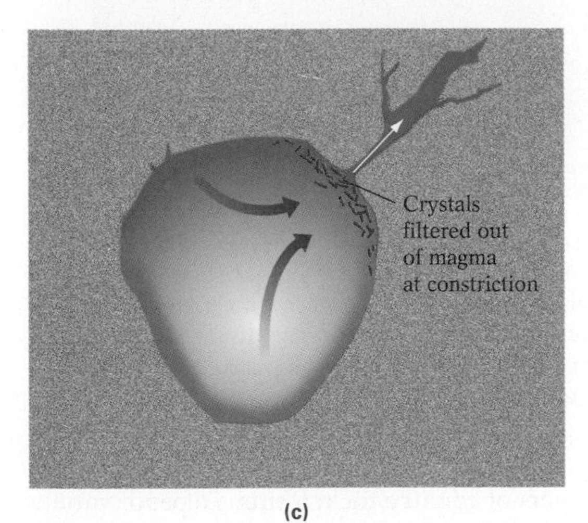

(c)

Figure 3-11 Early-forming crystals. Such crystals do not always remain in contact with the liquid magma, as Bowen's reaction series assumes. Instead, the crystals may **(a)** settle to the bottom of the magma chamber, **(b)** become affixed to the walls and roof of the magma chamber as magma circulates within the chamber, or **(c)** be filtered out of the magma as it is pressed through small fractures in the surrounding rock.

condition in which early-forming crystals remain in contact with the liquid magma, enabling them to interact and evolve continually until crystallization is complete. In nature, however, this condition rarely applies. As magma cools, some crystals may remain suspended, continuing to exchange ions and atoms with the remaining liquid. On the other hand, early-forming crystals might also be physically removed from the magma and have no further chemical interactions with the remaining liquid (Fig. 3-11). Crystals that are denser than the surrounding liquid may sink to the bottom of the magma chamber and become buried under later-settling crystals. In addition, crystals may be plastered against the walls or ceiling of the magma chamber by the hot rising liquid. The largest crystals may even be filtered out of the melt entirely, as the remaining liquid component of the magma flows into fractures too narrow to allow them to pass. Because their ions are no longer available to interact with the magma, removal of crystals in any of these ways limits the types of minerals that may later crystallize from the magma, thereby affecting the composition of any rocks that may form from it.

A magma from which crystals have been removed at various stages of its cooling has, in effect, become separated into a number of independently crystallizing bodies; the rocks that form from such a magma differ in composition both from each other and from the original magma. Because the silica-rich minerals are the last to crystallize, each successive body crystallized is more silica-rich (more felsic) than the last. By this process, called **fractional crystallization,** a single parent magma can produce a variety of igneous rocks of different compositions. The Palisades cliffs of northern New Jersey, on the west bank of the Hudson River, are a classic example of this phenomenon (Fig. 3-12).

Processes other than fractional crystallization can also significantly alter the composition of a moving magma. Blocks of rock from the walls of the magma chamber may break free, melt or partially melt, and become assimilated into the magma. In addition, two or more different bodies of magma may flow together and form a magma of hybrid composition. The 1912 volcanic eruption in Alaska's Aleutian Islands produced rocks containing both felsic and mafic minerals, suggesting that two distinct bodies of magma had combined to fuel the eruption.

Intrusive Rock Structures

Magmas tend to rise, for several reasons. When two materials of different densities occupy a space together, gravity pulls the denser down and the lighter is forced upward. Just as cream rises within nonhomogenized milk, magma rises because it is less dense than the solid rock surrounding it.

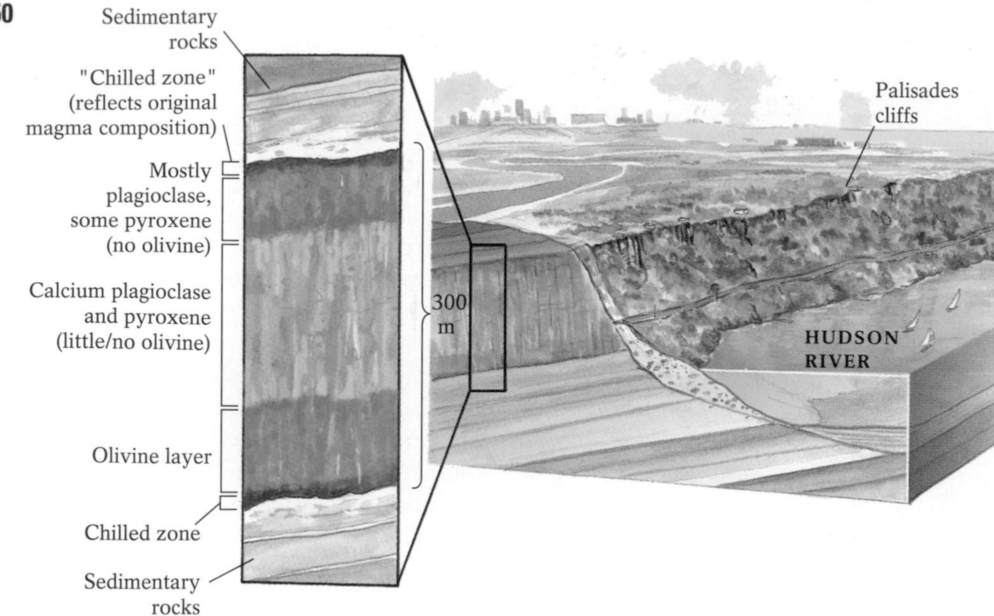

Figure 3-12 The New Jersey Palisades. This line of cliffs in the northeastern part of the state demonstrates the result of fractional crystallization. The rocks of the Palisades crystallized from a 300-meter (1000-foot)-thick body of magma that intruded preexisting rocks at temperatures of at least 1200°C (2200°F). The top and bottom of the Palisades solidified very rapidly without undergoing fractional crystallization, probably because the magma came into contact with cold surrounding rocks; they therefore provide us with a glimpse of the magma's original composition. The bottom third of the Palisades has a high concentration of olivine crystals, the central third is a mixture of calcium plagioclase and pyroxene with no appreciable olivine, and the upper third consists largely of plagioclase with no olivine and little pyroxene. Early-forming olivine apparently crystallized and then settled to the bottom of the magma body; pyroxene and plagioclase crystallized next, with the denser pyroxenes settling and concentrating in the center, and the lighter plagioclases occupying the uppermost section. Because the entire Palisades magma cooled and solidified fairly quickly, it left no residual magma from which later-forming minerals could crystallize.

In addition, because the pressure on a magma decreases with decreasing depth, the gases in the magma expand as it rises, helping to drive the material upward. Finally, magma rises when surrounding rocks press on it and squeeze it upward, much as toothpaste oozes out when you squeeze the tube.

As magma rises, it moves forcefully into cracks in preexisting rocks, pushing the rock aside to create its own space. Sometimes it may force overlying rocks to bulge upward, creating a domed intrusion within other rocks, known as a *diapir* (pronounced "DIE-uh-peer"). When moving magma incorporates some preexisting rock as it rises, some of the incorporated rocks may melt and become assimilated into the magma; Other rocks may remain unmelted and be carried within the magma. When this magma eventually solidifies, such "foreign" rocks appear as distinctly different rock masses called *xenoliths* (Fig. 3-13).

Figure 3-13 A dioritic xenolith within granite. The granitic magma encompassed the preexisting dioritic rock seen here as a dark gray mass within the lighter gray granite) but was not hot enough to melt it; as a result, the diorite was preserved as a discrete mass when the magma eventually solidified.

As magma cools and slowly crystallizes underground, it produces distinctive bodies of igneous rock that are referred to as **plutons.** These sometimes dramatic structures are often exposed at the surface after erosion removes the overlying rocks—in some areas, erosion has exposed thousands of kilometers of solidified intrusive magma. Plutons may be classified by their position relative to the preexisting rock, called *country rock,* surrounding it: *Concordant* plutons lie parallel to layers of country rock; *discordant* plutons cut across layers of country rock. Plutons of both varieties come in a range of shapes and sizes, as shown in Figure 3-14.

Tabular Plutons

Tabular plutons are slablike intrusions of igneous rock that are broader than they are thick, much like a table top. If magma flows into a relatively thin fracture in country rock or pushes between sedimentary rock layers, it will form a tabular pluton

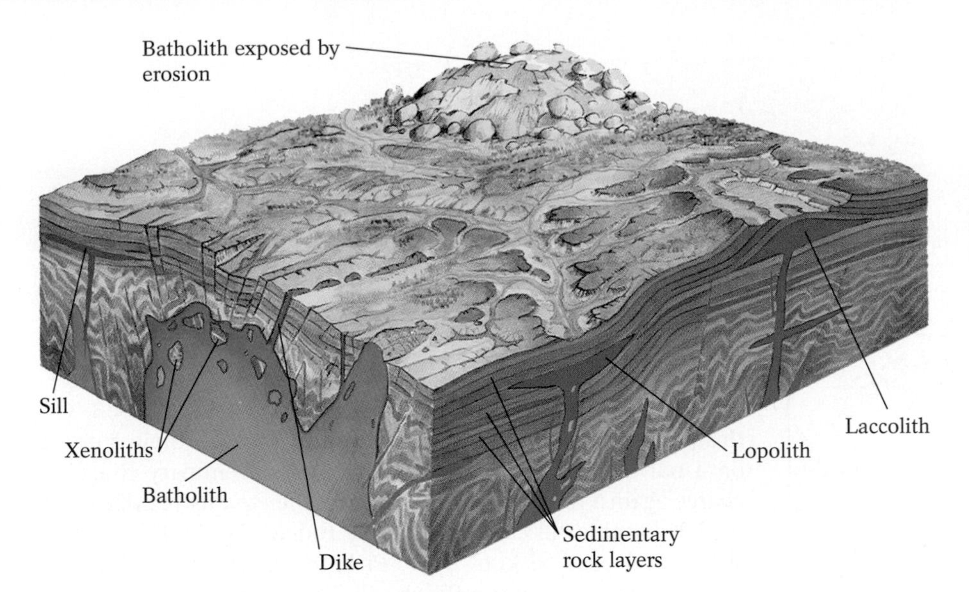

Figure 3-14 Plutonic igneous features. Sills are concordant tabular plutons; dikes are discordant tabular plutons. Laccoliths and lopoliths are larger concordant plutons, and batholiths are even larger discordant plutons.

when it cools. Tabular plutons may be as small as a few centimeters thick or as large as several hundred meters thick.

A **dike** is a discordant tabular pluton cutting across preexisting rocks. Dikes are generally steeply inclined or nearly vertical, suggesting that they formed from rising magma, which tends to follow the most direct route upward. These types of plutons often occur in clusters where magma infiltrated and solidified in a network of fractures.

Many dikes are created when magma rises into volcanoes and then solidifies; we can see these rocks when the less erosion-resistant material surrounding them wears away. Some dikes diverge like the spokes of a bicycle wheel from a *volcanic neck,* a vertical pluton remaining in what was once a volcano's central magma pathway. Along Route 64, running through the Navajo and Hopi lands of the Four Corners (the intersection of Colorado, New Mexico, Arizona, and Utah), we can see more than 100 such volcanic necks, remnants of ancient volcanic plumbing (Fig. 3-15).

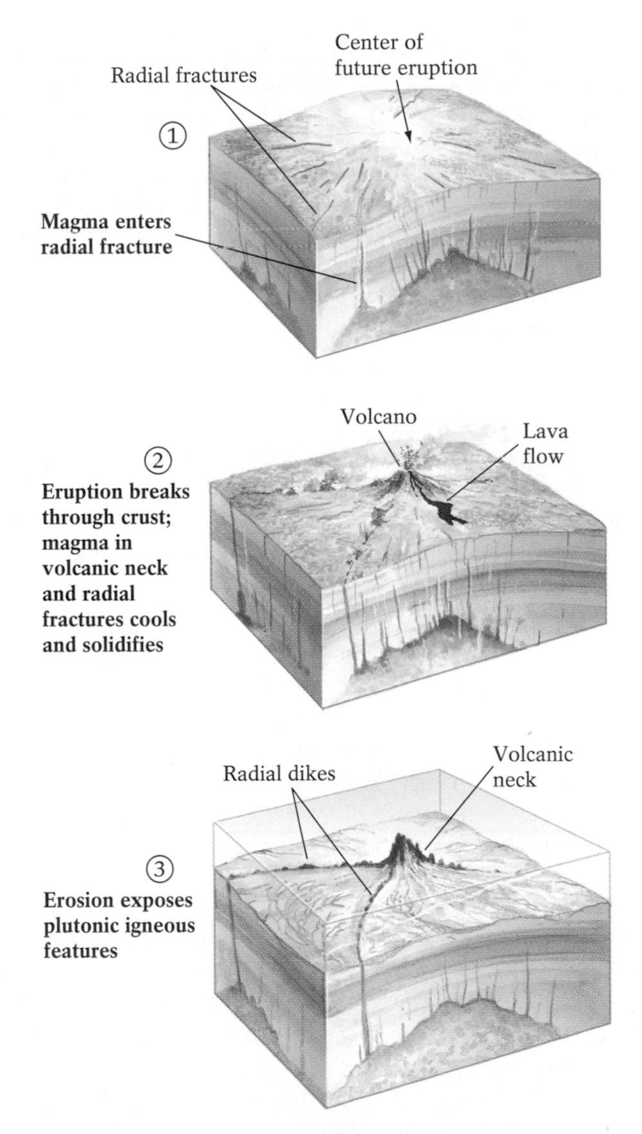

Figure 3-15 Shiprock Peak in New Mexico. This structure is believed to be a volcanic neck, the congealed lava from the interior of a former volcanic cone. Erosion of the surrounding sedimentary rock and the cone itself has exposed this volcanic neck and radial dikes.

Highlight 3-1 Tabular Plutons Save the Union

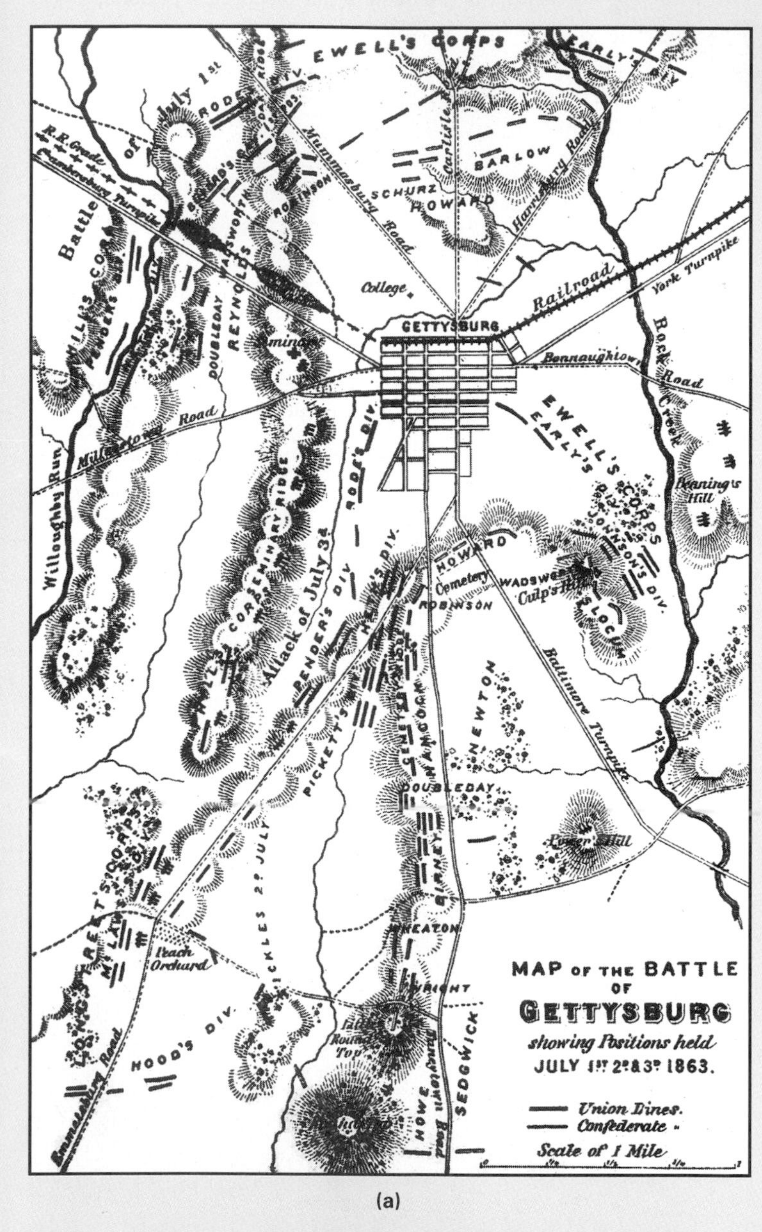

(a)

Figure 1 Geology and the Battle of Gettysburg. **(a)** A Civil War-era map of the site of the Battle of Gettysburg. Seminary Ridge appears in the upper left; Cemetery Ridge is to its lower right. **(b)** A contemporary artist's rendering of the relevant topographic features.

The Battle of Gettysburg, which lasted three days and took the lives of tens of thousands of Civil War soldiers, was effectively won by the Union on a hot July 3 in 1863. On this day, Confederate troops ventured forth from their outpost on a narrow dike of resistant basalt called Seminary Ridge to charge against the Union stronghold on the equally resistant, but thicker basaltic sill called Cemetery Ridge (Fig. 1). (This offensive would become known as "Pickett's charge.") The forward slope of the Cemetery Ridge sill impeded the Confederate charge, and a protective wall constructed from basaltic boulders by Union troops concealed them and repelled Confederate shots. Thus, with an assist from a well-placed basaltic sill, Union forces defeated the Confederate offensive at Gettysburg, a turning point in the American Civil War.

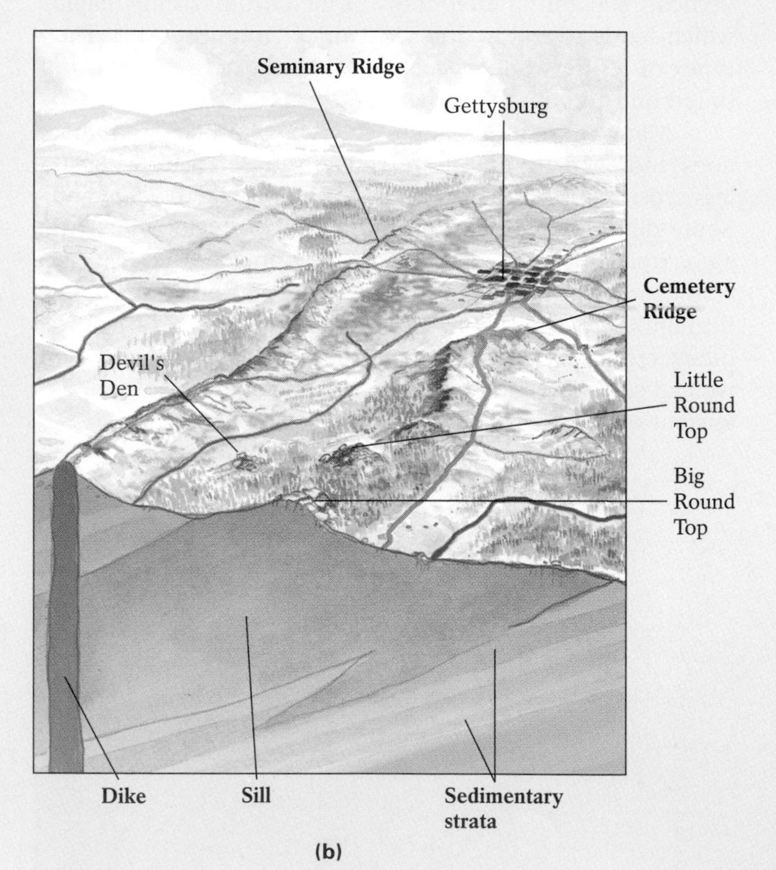

(b)

A **sill** is a concordant tabular pluton lying parallel to layers of preexisting rocks. Sills are produced when intruding magma enters a space between layers of rock, melting and incorporating adjacent sedimentary material. Sills can form only within a few kilometers of the Earth's surface, however—at greater depths, overlying rocks would compress and close off any spaces into which magma might flow. A sill and a dike in the south-central Pennsylvania town of Gettysburg provided the setting for an event that affected the course of American history, as recounted in Highlight 3-1.

Batholiths and Other Large Plutons

Large concordant plutons are commonly several kilometers thick and tens or even hundreds of kilometers across. They may be mushroom-shaped or saucer-shaped, close to the surface or deep beneath it. When thick, viscous magma intrudes between two parallel layers of rock and lifts the overlying one, it eventually cools to form a mushroom-shaped or domed concordant pluton, or **laccolith** (from the Greek *lakkos,* meaning "reservoir") (see Fig. 3-14). Laccoliths tend to form at relatively shallow depths, where little pressure acts to keep the overlying rock in place. They are typically granitic, formed from felsic magma that flows so slowly that it tends to bulge upward instead of spreading outward, raising the overlying rock to form a dome; erosion of this overlying dome may later expose the igneous rock below (Fig. 3-16). Sills form in a similar way but are usually basaltic and relatively flat, because they form from faster-flowing mafic magmas that can enter small spaces readily.

Unlike upward-bulging laccoliths, saucer-shaped concordant plutons called **lopoliths** (from the Greek *lopas,* meaning "saucer") sag downward (see Fig. 3-14). They are probably produced when dense mafic magma sinks as it intrudes, depressing the country rocks below to create a magma-filled basin. One such structure is evident along the western shore of Lake Superior, where the surrounding country rock has been eroded away (see Fig. 3-1).

Some igneous intrusions are even more vast than these large structures. **Batholiths** (from the Greek *bathos,* meaning "deep") are massive discordant plutons with surface areas (when exposed) of 100 square kilometers (40 square miles) or more (see Fig. 3-14). Most batholiths form at a depth of about 30 kilometers (20 miles) deep and are shaped somewhat like a human tooth—reaching their widest point at depth and then tapering to a point. They are generally found in elongated mountain ranges where erosion of overlying rocks has exposed deep cores of plutonic rocks, as in the White Mountains of New Hampshire and the Yosemite Valley of California's Sierra Nevada (Fig. 3-17).

Plate Tectonics and Igneous Rock

The worldwide distribution of igneous rock is not random. Certain structures and compositional types are found consistently in some geological settings, but not in others. Plutonic structures, for example, tend to form at or near the boundaries of diverging or converging tectonic plates, where

Figure 3-16 A laccolith along Route 95, through the Henry Mountains of southeastern Utah. The laccolith is the massive gray structure in the background.

Figure 3-17 Dioritic rocks. Virtually all the rock exposed in Kings Canyon National Park is composed of only one rock type—diorite. The panorama of dioritic rock here is part of the Sierra Nevada batholith.

fracturing rock provides openings in which magma can intrude. Smaller plutonic features, such as dikes and sills, generally appear in divergent or rifting zones, where mafic magmas rise as the Earth's brittle outer layers become stretched and pulled apart. Intermediate and granitic batholiths are found where oceanic plates have subducted, marking the sites of both modern and ancient plate boundaries. Oceanic rocks carried by subduction down into the asthenosphere are partially melted, generating the vast quantities of magma that form coastal batholiths. The chain of batholiths in western North America, which stretches from British Columbia through the California Sierras to Baja California, developed through more than 200 million years of oceanic-plate subduction.

Like igneous structures, the compositions of the common igneous rocks—mafic through felsic—are associated with specific geological settings. Most igneous rocks form where the three factors that create magma—heat, reduced pressure, and water (see page 47)—come into play most dramatically: at active plate boundaries. The type of rocks produced depends largely on the type of plate boundaries involved.

Basalts and Gabbros

The mafic volcanic rock basalt and its plutonic equivalent gabbro are the only igneous rocks found in oceanic crust. Basalts are also relatively common in continental crust. The low viscosity of mafic magma allows much of it to flow to the Earth's surface, where it erupts as basalt. Consequently, gabbros are rarely seen at the Earth's surface, and most of what we know about the origin of mafic rock comes from basalts. Basalts are the most abundant and most variable igneous rocks, occurring in a number of different tectonic settings and in a range of compositions. The compositions of the various basalts apparently depend on whether their parent magmas derived from deep- or shallow-mantle sources (Fig. 3-18).

Mid-ocean ridge basalts are the most abundant volcanic rock, accounting for 65% of the Earth's surface area. They are produced by eruptions at oceanic divergent boundaries and probably originate from partial melting of the upper mantle. *Ocean island basalts*—such as the ones that formed the Hawai'ian Islands—do not occur at divergent plate boundaries but rather appear at "hot spots," volcanic zones (generally intraplate) that overlie deep mantle sources.

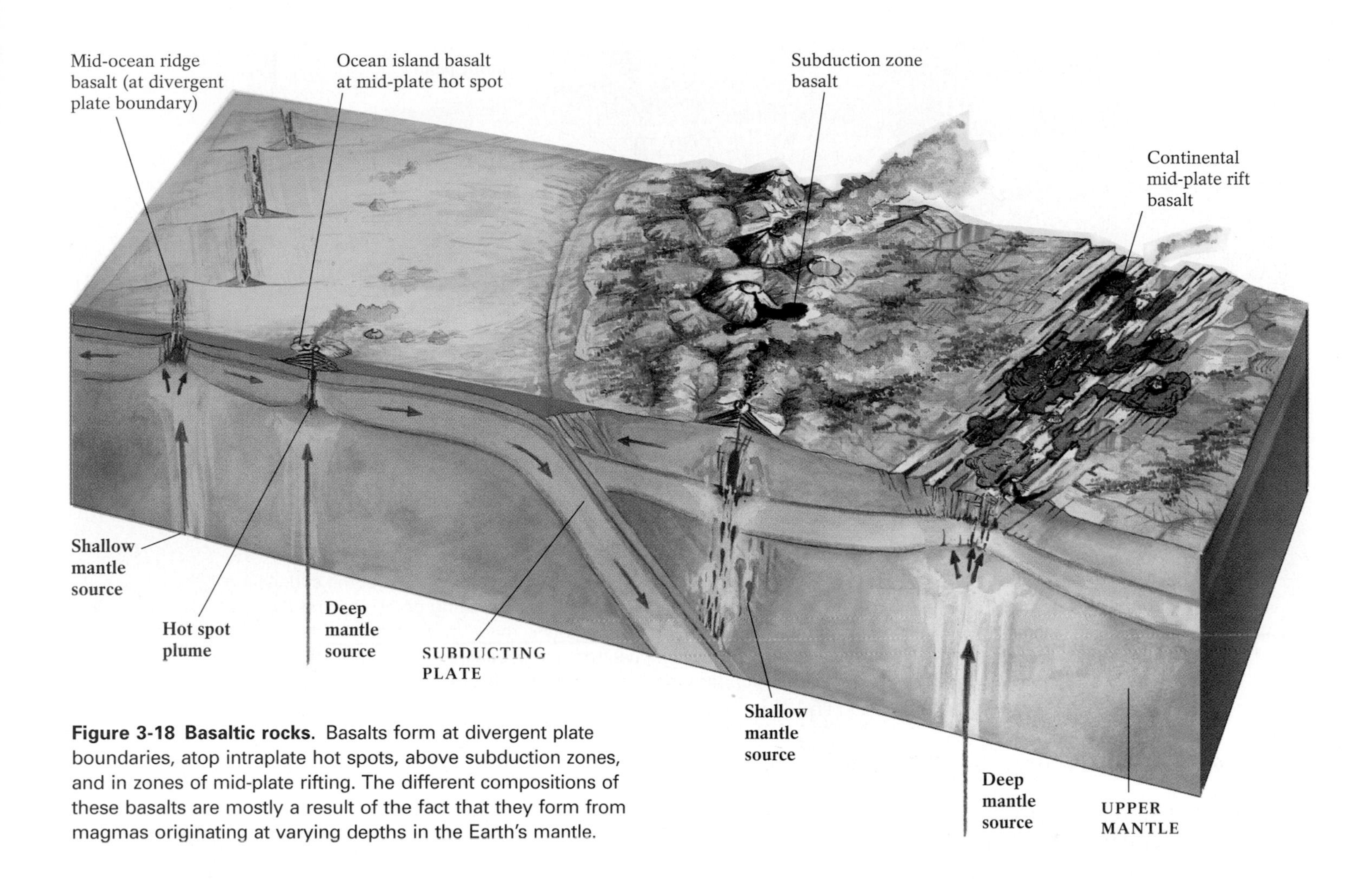

Figure 3-18 Basaltic rocks. Basalts form at divergent plate boundaries, atop intraplate hot spots, above subduction zones, and in zones of mid-plate rifting. The different compositions of these basalts are mostly a result of the fact that they form from magmas originating at varying depths in the Earth's mantle.

Basalts in continental settings vary more in composition than do oceanic basalts. Those associated with continental rifting are most likely derived from deep-mantle sources; others, which arise at subduction zones, tap shallower sources. Both types of basalt form, however, as hot mafic magma rises through tens of kilometers of continental crust, incorporating many of the materials in its path and gradually changing in composition.

Andesites and Diorites

The less mafic (intermediate) rocks andesite and diorite are commonly found along the subductive margins of continents and on volcanic islands formed through subduction of oceanic plates. Regions of andesitic rock are found on virtually all the lands that border the Pacific Ocean (see Fig. 3-8); this **andesite line** follows the nearly continuous ring of subduction zones surrounding the Pacific Ocean basin (Fig. 3-19).

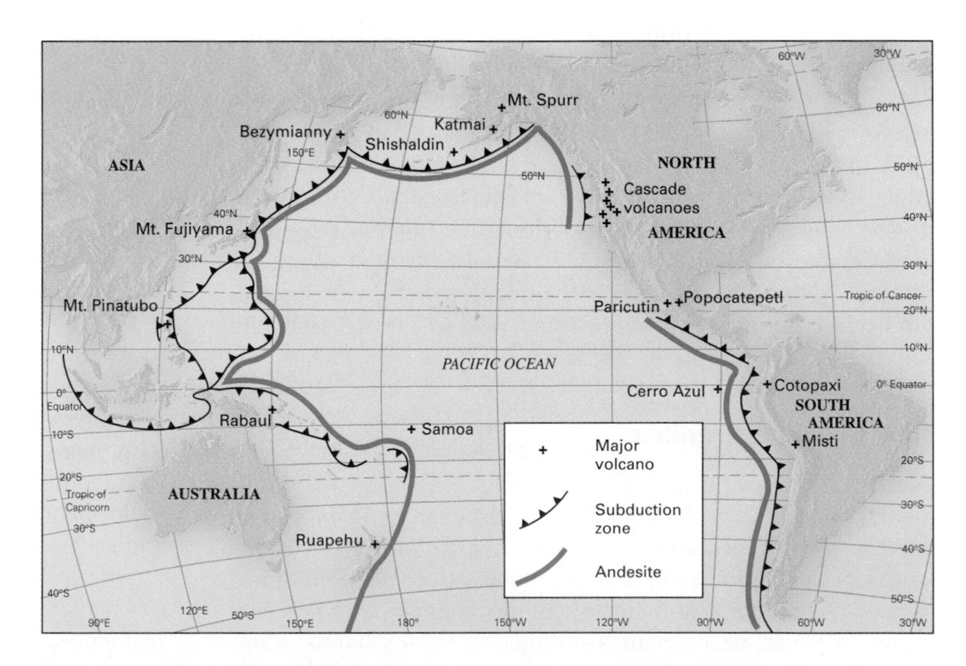

Figure 3-19 The andesite line. Subduction-produced andesitic and dioritic rocks make up most of the surface geology surrounding the Pacific Ocean basin.

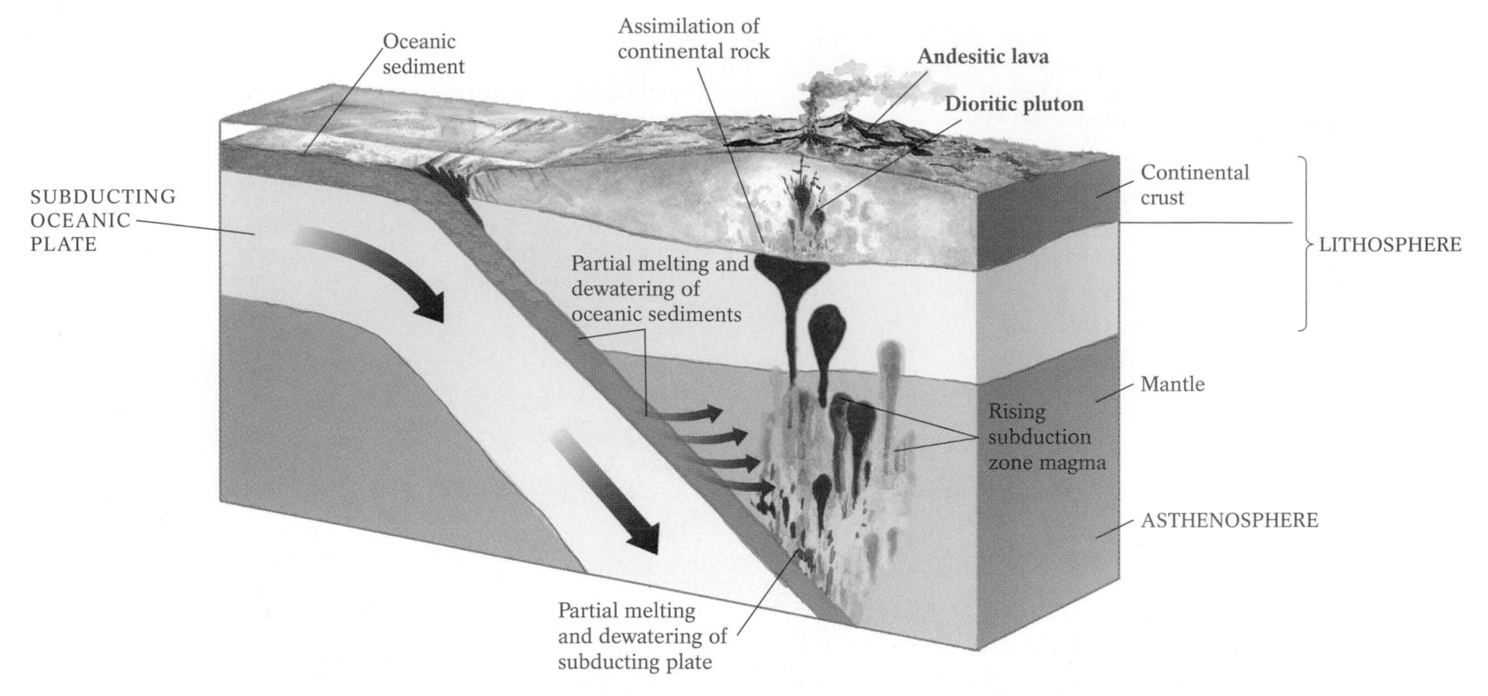

Figure 3-20 The factors involved in the origin of andesite and diorite. Water pressed out of the subducting plate and its associated oceanic sediment enters the mantle rock and lowers its melting point, causing it to melt and rise as basaltic magma. The composition of this initially mafic magma is made more intermediate by its mixing with partially melted felsic oceanic sediment and oceanic crust from the subducting plate, as well as with felsic country rock assimilated by the magma as it rises up through the continental crust of the non-subducting plate.

A number of processes are believed to combine to produce rocks of intermediate composition from subducting oceanic plates, beginning with the production of an initially mafic magma. As an oceanic plate subducts, increasing pressure drives water from the plate and forces its overlying sediments into the ultramafic mantle. The water lowers the melting point of rocks within the mantle (see page 47), producing a basaltic magma. As this magma rises, it melts and assimilates a portion of the mostly felsic oceanic sediments—averaging 200 meters (650 feet) thick on the ocean floor—carried on the subducting plate. Partial melting of the subducting plate itself may also contribute an intermediate component to the magma. Finally, a subduction-zone magma may melt and assimilate felsic materials as it rises through overlying continental rocks. Figure 3-20 summarizes the various factors that combine to produce andesite and diorite from the subduction of oceanic lithosphere.

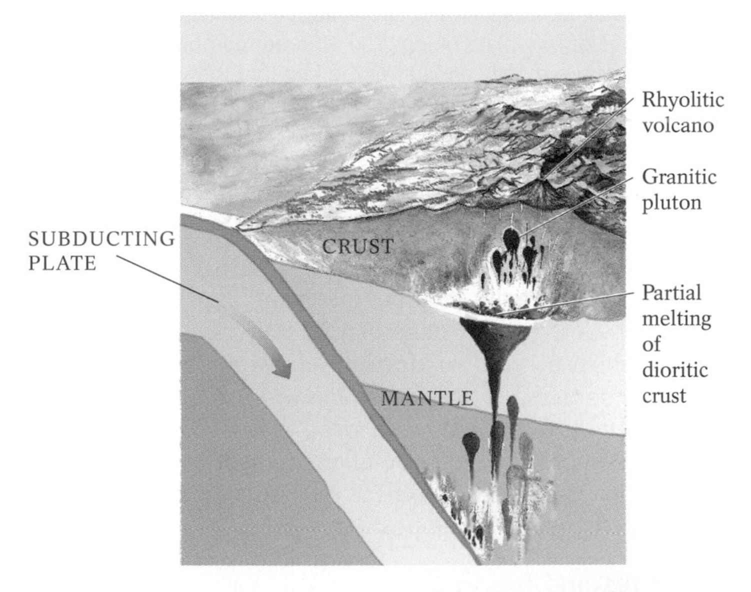

Figure 3-21 The origin of felsic rocks at subducting-plate boundaries. Hot rising mafic and intermediate magmas partially melt dioritic rocks in the lower continental crust, producing granitic plutons (and, occasionally, granite's rare volcanic equivalent, rhyolite).

Rhyolites and Granites

Nearly all rhyolitic and granitic rocks are found on continents, and they probably originate principally from partial melting of lower continental crust. Most granitic intrusions appear near modern or ancient subduction plate margins, where rising mafic and intermediate magmas and the frictional heat that accompanies subduction caused partial melting of dioritic rocks at the base of plate-edge mountain belts (Fig. 3-21). Partial melting of intermediate rocks produces predominantly felsic magma. Because these magmas are typically very viscous, they tend to rise slowly and cool at depth, producing the felsic plutonic rock granite. When such magmas reach the surface (as they sometimes do when their water content is high), they erupt as rhyolite.

Igneous Rocks on the Moon

Since 1969, geologists who study igneous rocks have extended the reach of their rock hammers some 400,000 kilometers (240,000 miles) to the Moon. The Moon appears to be fundamentally different from the Earth in terms of its geology. First, although recent studies suggest that water may actually exist in isolated, protected spots near the Moon's poles, for all practical purposes, the Moon is completely waterless. If any substantial amount of water ever existed on the Moon, it became heated, vaporized, and escaped the Moon's gravitational field very early in the Moon's evolution. Second, the origin of the Moon's igneous rocks seems unrelated to plate tectonics—most lunar geologists believe that the Moon has never had moving plates. Likewise, the Moon's igneous activity seems unrelated to its internal heat, much of which dissipated long ago.

Moon rocks collected by Apollo astronauts in the 1970s indicate that the Moon's surface contains at least two distinct types of geological/geographical provinces—the highlands and the *maria* (pronounced "MAR-ee-a"; plural of Latin *mare,* meaning "sea"). The lunar highlands probably crystallized as the Moon's earliest crust some 4.5 to 4.0 billion years ago, when the early Moon's interior was hot enough to develop a multilayered structure in a process similar to the one that created the Earth's interior. Its rocks consist principally of anorthosite, a type of coarse-grained plutonic igneous rock composed almost exclusively of the calcium plagioclase mineral anorthite.

The lunar maria, or "seas," are actually vast solidified basaltic flows (Fig. 3-22). (Galileo and other early astronomers, who used the crude telescopes of the time, believed that they were true seas—hence their name.) The maria

(a)

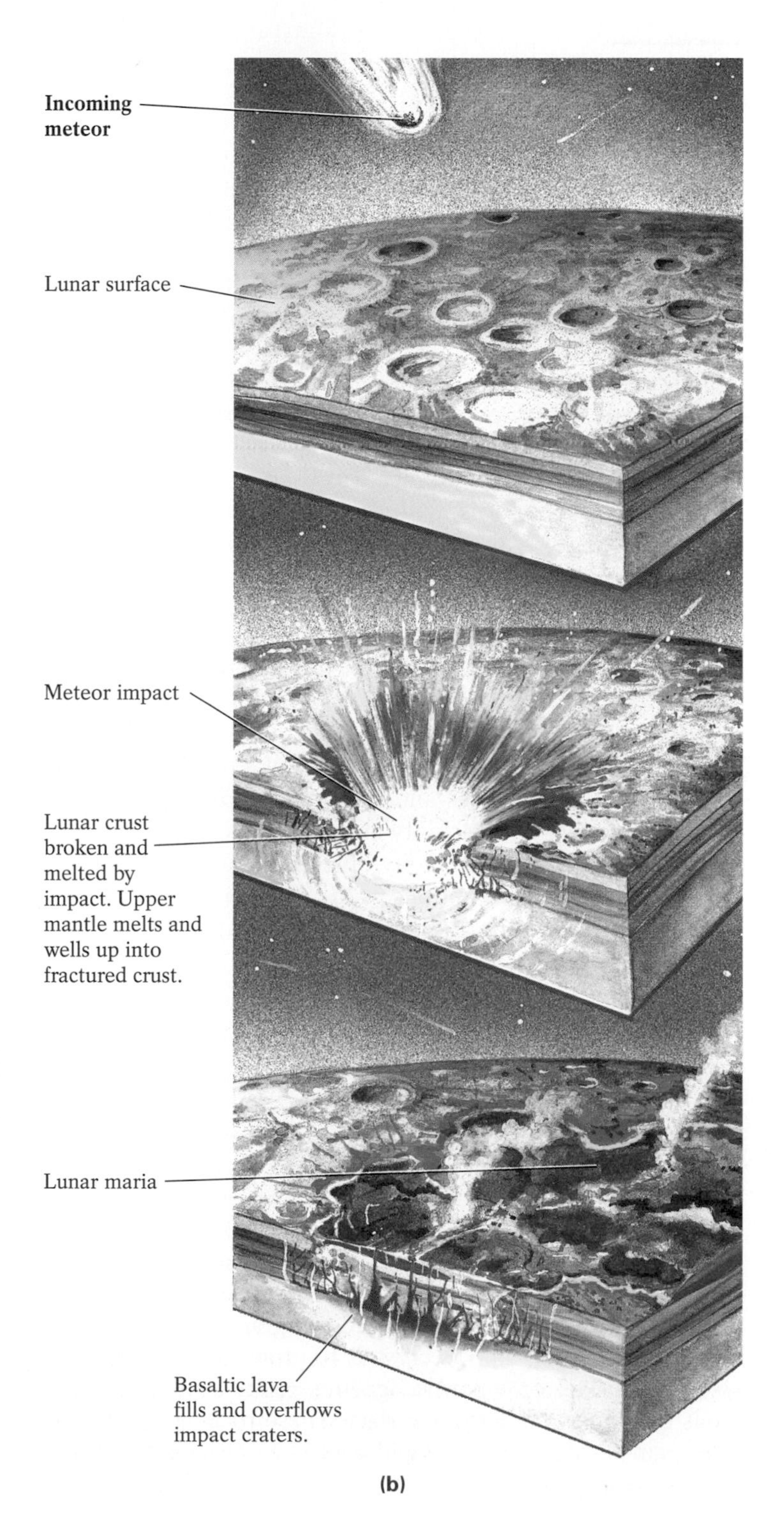

(b)

Figure 3-22 Geological provinces of the Moon. (a) A telescopic view of the near side of the Moon, showing its highlands and maria. **(b)** The formation of lunar maria from meteor impacts.

probably formed 4 to 3.85 billion years ago, when intense meteorite activity gouged numerous craters in the Moon's surface. The impacts fractured the lunar crust, providing subsurface magmas with easy pathways to the surface; they may also have raised the temperature of the stricken rocks to their melting points, thus generating new magma. Basaltic lava flowed into and filled the craters, forming the lunar maria.

The Economic Value of Igneous Rocks

The practical uses of igneous materials range from the glittering (gemstones and precious metals) to the utilitarian (crushed basalt for road construction). Any urban center displays one of the principal applications for plutonic igneous rock—the decorative building stone that adorns the exteriors and lobbies of many banks and office buildings. The same appealing polished granites and diorites can be found in cemeteries, where they serve as durable tombstones. On a smaller scale, glassy pumice serves as the abrasive in grease-removing cleansers and is used to remove calluses from hands and feet. Until recently, pumice was used as an ingredient in toothpaste, thanks to its ability to remove dental stains and plaque; because it also claimed its share of tooth enamel, however, it has since been replaced by milder abrasives. Some other familiar and useful minerals, such as the diamonds found in ultramafic rocks and the emeralds and topazes in felsic pegmatitic rocks, are also of igneous origin. Gold and silver are often found in or around granitic rocks, as are the less shiny but still valuable ores of copper, lead, and zinc.

We can now build on our general knowledge of the Earth's igneous processes and rocks and expand our discussion to cover igneous activity occurring above ground. The next chapter focuses on readily observable igneous phenomena—the volcanic eruptions and rocks produced when magma reaches the Earth's surface and escapes into the air.

Chapter Summary

Igneous rocks, the most abundant type of rock in the Earth's crust and mantle, form when molten rock cools and crystallizes. **Magma** is molten rock that flows within the Earth. When magma reaches the Earth's surface, it becomes **lava,** molten rock that flows above ground. **Intrusive,** or **plutonic,** igneous rocks form from magma that cools slowly underground. These rocks are generally coarse-grained, because their ample cooling time allows crystals to grow to relatively large sizes. **Extrusive,** or **volcanic,** igneous rocks form when lava cools quickly at the Earth's surface. These rocks are generally fine-grained, because their rapid cooling limits crystal growth.

The most common igneous rocks include ultramafic **peridotite** (containing less than 40% silica), mafic **basalt** and **gabbro** (40%–55% silica), intermediate **andesite** and **diorite** (55%–65% silica), and felsic **granite** and **rhyolite** (65% or more silica).

Magmas are produced in the Earth's interior when preexisting rocks undergo **partial melting**—that is, the minerals with lower melting points liquefy first and start to flow as a molten mass; this mass includes still-solid crystals of minerals that melt at higher temperatures. Other factors that influence the creation of magma include heat, pressure, and water content.

As magma cools, different minerals crystallize from it at different temperatures. The silicate minerals, as a group, crystallize in two specific sequences, known collectively as **Bowen's reaction series.** In the discontinuous series, the ferromagnesian (iron- and aluminum-rich) minerals evolve in distinct steps, with both their composition and their internal crystal structures changing at each step. In the continuous series, the plagioclase feldspars evolve as sodium ions gradually replace calcium ions in the developing crystals, without any accompanying change in the minerals' internal crystal structures.

The removal of early-forming crystals from a magma means that their ions are no longer available to interact with the magma, limiting the types of minerals that may later crystallize from it. As a result of this process, called **fractional crystallization,** the rocks ultimately produced by the magma will differ in composition from those that would have been produced by the original, unseparated magma. The composition of a magma may also be modified by assimilation of preexisting rocks or by mixing with another body of magma having a different composition.

Bodies of magma that cool underground form **plutons,** igneous structures that are distinct from the surrounding rocks. Plutons are classified by their shapes, sizes, and orientation relative to the rocks they intrude. Concordant plutons occur parallel to the preexisting rock layers; discordant plutons cut across the preexisting rock layers. Tabular plutons are relatively thin, igneous structures, much like a tabletop. Discordant tabular plutons are called **dikes**; concordant tabular plutons are called **sills.** Large concordant plutons include mushroom-shaped **laccoliths** and saucer-shaped **lopoliths**; large discordant plutons are called **batholiths.**

The principal igneous rock types are typically associated with specific plate tectonic settings. Basalts and gabbros are found at oceanic divergent boundaries (the mid-ocean ridge basalts), atop intraplate hot spots (the ocean island basalts), at continental margins where oceanic plates have subducted, and near rifting continental plates. Andesites and diorites appear where oceanic plates have subducted, both at continental margins and on volcanic islands. The nearly continuous ring of subduction-produced andesites that surrounds the Pacific Ocean is called the **andesite line.** Rhyolites and granites are formed within continents by partial melting of the lower portions of the continental crust; these types of igneous rocks are often associated with subduction-produced mountains.

Igneous rocks are also found on the Moon. Lunar igneous rocks differ fundamentally from those on Earth, in that they contain no water and their formation involved neither plate tectonics nor subsurface heat. The Moon's surface consists of highlands composed largely of anorthosite, a coarse-grained plutonic igneous rock, and vast areas of basalt known as maria.

Igneous rocks are valued for the gemstones and precious metals they contain. They are also used for a variety of practical purposes, such as road construction, architectural design, and household abrasives.

Key Terms

igneous rocks (p. 41)
magma (p. 42)
lava (p. 42)
intrusive rocks (p. 43)
plutonic rocks (p. 43)
extrusive rocks (p. 43)
volcanic rocks (p. 43)
peridotite (p. 45)
basalt (p. 45)
gabbro (p. 46)
andesite (p. 46)
diorite (p. 46)
granite (p. 46)
rhyolite (p. 46)
partial melting (p. 46)
Bowen's reaction series (p. 48)
fractional crystallization (p. 49)
plutons (p. 50)
dike (p. 51)
sill (p. 53)
laccolith (p. 53)
lopoliths (p. 53)
batholiths (p. 53)
andesite line (p. 55)

Questions for Review

1. Briefly describe the textural difference between phaneritic and aphanitic rocks. Why do these rocks have different textures?

2. Some igneous rocks contain large visible crystals surrounded by microscopically small crystals. What are these rocks called? How does such a texture form?

3. What elements would you expect to predominate in a mafic igneous rock? In a felsic igneous rock?

4. Name the common *extrusive* igneous rocks in which you would expect to find each of the following mineral types: calcium feldspar; potassium feldspar; muscovite mica; olivine; amphiboles; sodium feldspars. Which *plutonic* igneous rock contains abundant quartz and muscovite mica, but virtually no olivine or pyroxene?

5. What factors, in addition to heat, control the melting of rocks to generate magma?

6. What is the basic difference between the continuous and discontinuous series of Bowen's reaction series?

7. Briefly describe three things that might happen to an early-crystallized mineral surrounded by liquid magma.

8. How do a sill and a dike differ? A laccolith and a lopolith? A lopolith and a batholith?

9. Briefly discuss two specific types of plate tectonic boundaries and the igneous rocks that are associated with them.

10. What is the basic difference between a mid-ocean ridge basalt and an oceanic island basalt?

For Further Thought

1. What type of igneous feature is shown in the photo below?

2. Felsic rocks such as rhyolite often occur together with basaltic rocks near rifting continents. Give one possible explanation for this pairing.

3. Why do we rarely find batholiths made of gabbro?

4. How might the distribution of the Earth's igneous rocks change when the Earth's internal heat is exhausted and plate tectonic movement stops?

5. Why are there virtually no granites or diorites on the Moon? How might small volumes of such felsic rock form under the geological conditions believed to be responsible for the Moon's igneous rocks?

4

Volcanoes and Volcanism

One summer day in 1883, the Indonesian island of Krakatoa all but disappeared in a spectacular, massive volcanic eruption. Krakatoa, an uninhabited volcanic island in the Sunda Straits of the southwest Pacific Ocean, had for many years served as a landmark for clipper ships carrying tea from China to England. The volcano, which had been inactive for more than two hundred years, stood 792.5 meters (2601 feet) high. On the morning of August 27, it suddenly erupted in one of modern history's most violent volcanic eruptions, with a force equivalent to the explosion of 100 million tons of TNT. No one, as far as we know, perished directly from the destruction of Krakatoa; however, between 36,000 and 100,000 lives were lost as the resulting ocean waves, up to 37 meters (121 feet) high, pounded coastal villages on the nearby islands of Java and Sumatra. The sound of the explosion was heard in places as distant as central Australia, 4802 kilometers (2983 miles) away, which is akin to the residents of San Diego, California, hearing an explosion in Boston, Massachusetts. The eruption produced a black cloud of volcanic debris that rose to an altitude of 80 kilometers (50 miles), blocked out all sunlight, and plunged the region into darkness for three days. The cloud's finest particles, swept aloft by wind currents, reduced incoming solar radiation by as much as 10% worldwide, causing a drop of more than 1°C (1.8°F) in global temperatures and leading to years of spectacular crimson sunsets.

Volcanoes are the landforms created when molten rock escapes from the Earth's interior through openings, or *vents*, in the Earth's surface and then cools and solidifies around the vents. An estimated 600 volcanoes have erupted in the past 2000 years, some of them many times over. In a single year, approximately 50 volcanoes erupt around the world (Fig. 4-1). As the Krakatoa example illustrates, powerful volcanic eruptions and their aftereffects can be among the Earth's most destructive natural events. On the other hand, volcanoes also provide some of the world's most breathtaking scenery. Each year, millions are drawn to the slopes of Mount Rainier in Washington state, Mount Fuji in Japan, and Mount Vesuvius in Italy. Volcanic activity also adds to the

Figure 4-1 Popocatepetl ("El Popo"). Mexico City's neighboring volcano is one of several volcanoes currently erupting around the world and threatening surrounding communities.

Figure 4-2 Anak Krakatau ("Child of Krakatoa"), the small island that emerged from the remains of the volcanic island Krakatoa during eruptions in the 1920s. The original Krakatoa volcano was demolished in a monumental eruption in 1883.

Earth's inventory of habitable real estate: Iceland, Japan, Hawai'i, Tahiti, many islands of the Pacific and Caribbean, and nearly all of Central America are products of volcanic activity. A new volcanic island is even growing where Krakatoa once stood (Fig. 4-2).

The geological processes that result in the expulsion of molten rock as lava at the Earth's surface are collectively known as **volcanism.** Volcanoes and volcanism are simultaneously a great hazard and a great boon to humankind. In this chapter, we explore the causes and characteristics of the different types of volcanoes. We describe the threats they pose and our strategies for coping with them. Finally, we examine volcanism on some of our neighboring planets.

The Nature and Origin of Volcanoes

Volcanoes and other, less dramatic manifestations of volcanism offer windows into the Earth, providing us with information and materials that would otherwise remain inaccessible. Ascending magma carries subterranean rock fragments to the surface, giving us a glimpse of actual rocks from the Earth's interior. We owe the air we breathe and much of the Earth's water to volcanic eruptions, which have released useful gases from the Earth's interior throughout the planet's existence. Volcanic terrains often become prime agricultural lands as fresh volcanic ash replenishes the nutrients in nearby soils. Volcanism can even be a source of inexpensive, clean energy. Iceland, for example, has an abundant underground hot water supply, thanks to the molten rock that fuels its volcanic activity; by tapping the scalding water just meters beneath their feet, Icelanders can heat more than 80% of their homes and businesses.

Volcano Status

Whether a volcano poses an imminent threat to human life and property depends on its current status as either an active, dormant, or extinct volcano. An *active* volcano is one that is currently erupting or has erupted recently (in geological terms). Certain active volcanoes, such as K'ilauea on Hawai'i or Stromboli in the eastern Mediterranean, erupt almost continuously. Others erupt periodically, such as Lassen Peak in northern California, whose last activity occurred in 1917. Active volcanoes can be found on all continents except Australia and on the floors of all the major ocean basins. Indonesia, with 76 active volcanoes, Japan, with 60, and the United States, with 53, are the world's most volcanically active nations.

A *dormant* volcano is one that has not erupted recently (within the past few thousand years) but is considered likely to do so in the future. The presence of relatively fresh (less than 1000 years old) volcanic rocks in a volcano's vicinity suggests that it is still capable of erupting (Fig. 4-3). Signs of rising magma, such as the presence of hot water springs or small earthquakes occurring near a volcano, may indicate that the volcano is stirring to wakefulness.

A volcano is considered *extinct* if it has not erupted for a very long time (perhaps for tens of thousands of years) and is considered unlikely to do so in the future. One indication that a volcano is probably extinct is that extensive erosion has taken place on its slopes since its last eruption. A truly extinct volcano is no longer fueled by a magma source. Volcanoes can, however, surprise us. Residents of the Icelandic island of Heimaey believed their Mount Helgafjell to be extinct until it came to life in a spectacular eruption in 1973, its first in 5000 years.

The Causes of Volcanism

Volcanism begins when magma created by the melting of preexisting rock (discussed in Chapter 3) travels through fractures in the lithosphere to reach the Earth's surface. The distribution of the Earth's lithospheric cracks, which are usually associated with tectonic plate boundaries and with intraplate hot spots, determines where most volcanoes will form. A magma will erupt if it flows upward rapidly enough to reach the surface before it can cool and solidify. Two characteristics of a magma determine its potential to achieve eruption: its gas content and its viscosity.

(a)

(b)

Figure 4-3 Which of these two volcanoes is more likely to erupt? (a) The slopes of Washington state's Mount Rainier are deeply scored by erosion and appear not to have received a fresh covering of lava in thousands of years. **(b)** The slopes of Mount St. Helens (shown here before its 1980 eruption) are relatively uneroded. Although intermittently dormant for hundreds or thousands of years, Mount St. Helens' volcanic cone has continued to grow.

Gas in Volcanic Magma Magmatic gases make up 1% to 9% of most magmas. The principal gases are water vapor and carbon dioxide, though smaller quantities of nitrogen, sulfur dioxide, chlorine, and a few others may also be present. Tens of kilometers underground, these gases remain dissolved in magma, trapped by the pressure of the surrounding rocks. As the magma rises toward the surface, the decreased pressure causes the gases to begin to leave the solution. The released gases, which are less dense than the surrounding magma, migrate upward, pushing any overlying magma before them. The higher the gas content of a magma, the faster it will rise and the greater its chances of reaching the surface before solidifying.

Gases become concentrated near the top of a rising magma body and press against the overlying rock. When these volcanic gases are completely prevented from escaping, perhaps by a plug of congealed lava blocking the passage to the surface, they accumulate and exert even greater pressure against the overlying rock. Ultimately, the overlying rock shatters. The pent-up gases then expand instantaneously, much as the gases in an agitated soft-drink bottle fizz and bubble out when you open the bottle. The initial blast removes any overlying obstructions, hurling masses of older rock skyward. Shreds of the liberated, gas-driven lava are sprayed violently into the air as the gases expand. The eruption may continue violently for hours or even days as the gases escape. The eruption may later settle down to a relatively placid outpouring of degassed magma, or it may cease altogether.

Magma Viscosity A magma's viscosity (resistance to flow) generally decreases with heat and increases with its silica content. Felsic magma tends to be relatively cool (because it crystallizes at low temperatures) and has a high silica content; thus it is very viscous. Conversely, because mafic magma is hot and has a low silica content, it is much less viscous and flows easily. For this reason, mafic magmas are more likely to rise to the surface and erupt than are felsic magmas, which tend to cool underground and form plutonic rocks.

The viscosity of magma has a direct effect on the explosiveness of a volcanic eruption. In more fluid, mafic magmas, rising gases meet with little resistance and therefore escape readily and relatively quickly when the magma reaches the surface. In highly viscous felsic magmas, the slower movement of gases (analogous to air bubbles that tend to rise more slowly in a thick milkshake than they do in water) allows gas pressures to build within the molten material. Thus felsic magmas tend to erupt explosively.

The Products of Volcanism

Volcanic eruptions range from the quiet oozing of basaltic lava, such as flows from K'ilauea volcano in Hawai'i, to Krakatoa-type cataclysmic explosions. Depending largely on the composition of the magma that feeds it, a volcanic eruption can produce a flowing stream of red-hot lava, a shower of ash particles as fine as talcum powder, a hail of volcanic blocks the size of automobiles, or any number of intermediate-sized products. In this section, we first examine the different types of lava flows and then describe the various forms in which volcanic material is deposited on the surface.

Types of Lava Flows

As noted earlier, because mafic magmas are extremely hot and relatively fluid, they are more likely to rise to the surface and erupt than are felsic magmas. Thus basaltic lava is the most common type of lava. The tendency of basaltic magma to become volcanic explains why we find much more basalt than gabbro (basalt's plutonic equivalent) in the Earth's crust. Magmas of felsic composition tend to be cooler and much more viscous, only rarely reaching the surface before solidifying. For this reason, crustal rocks contain much more granite than rhyolite. Andesitic lavas are intermediate between basaltic and rhyolitic lavas in both composition and fluidity; they erupt much more frequently than rhyolitic lava but are less common than basalt.

Basaltic Lava Much of what we know of basaltic lava comes from observing the nearly continuous eruptions on the Big Island of Hawai'i. The temperature of the Hawai'ian flows can reach as high as 1175°C (2150°F). Such hot, low-viscosity lava cools to produce two principal types of basalt, *pahoehoe* (pronounced "pa-HOY-hoy") and *'a'a* (pronounced "AH-ah"). Pahoehoe, which means "ropy" in a Polynesian dialect, is aptly named. Highly fluid basaltic lava moves swiftly down a steep slope at speeds that may exceed 30 kilometers (20 miles) per hour, spreading out rapidly into sheets about 1 meter (3.3 feet) thick. The surface of such a flow cools to form an elastic skin that is dragged into ropelike folds by the continuing movement of the still-fluid lava beneath. The ropy surface of pahoehoe is generally quite smooth (Fig. 4-4).

Figure 4-4 Pahoehoe lavas. These ropy lavas from Hawai'i's K'ilauea volcano cooled only days or even hours before these tourists began trekking around on them.

Figure 4-5 'A'a lavas. This relatively slow-moving basaltic lava cools to form a blocky, jagged, 'a'a-type surface texture.

Native islanders refer to it as "ground you can walk on barefoot," and most old Hawai'ian foot trails follow ancient pahoehoe flows.

As basaltic lava flows farther from its vent, it cools and becomes increasingly viscous. A thick brittle crust develops at its surface and slowly continues to move forward, carried along by the warmer, more fluid lava below it. The molten interior of the flow advances more rapidly than the cooler outer region, breaking it up to produce a rough surface having numerous jagged projections sharp enough to cut animals' hooves. Flows having these features are called 'a'a flows (Fig. 4-5). (*'a'a* is a local term of unknown origin that may recall the cries of a barefoot islander who strayed onto its surface; ancient foot trails meticulously avoid 'a'a fields.) 'A'a is often found downstream from pahoehoe, the product of the same flow.

Basaltic flows may produce several other distinctive features as they cool. For example, gas still present in the lava often migrates to its surface leaving small pea-sized voids, or *vesicles*, which may be preserved at the top of the basalt when it cools. Such vesicle-rich basalt is known as *scoria*. As basaltic lava cools and solidifies into rock, it often contracts in size, producing a polygonal pattern of cracks. The cracks extend from the top and bottom surfaces of the flow into its interior as cooling progresses, creating six-sided columns of rock (Fig. 4-6). In North America, the Devil's Postpile in California's Sierra Nevada and Devil's Tower in northeastern Wyoming (site of the climax of the film *Close Encounters of the Third Kind*) are spectacular examples of basaltic columns. Basaltic flows may also contain large *lava tubes*, which form when lava solidifies into a crust at its surface but continues to flow inside, forming a tunnel that is eventually drained

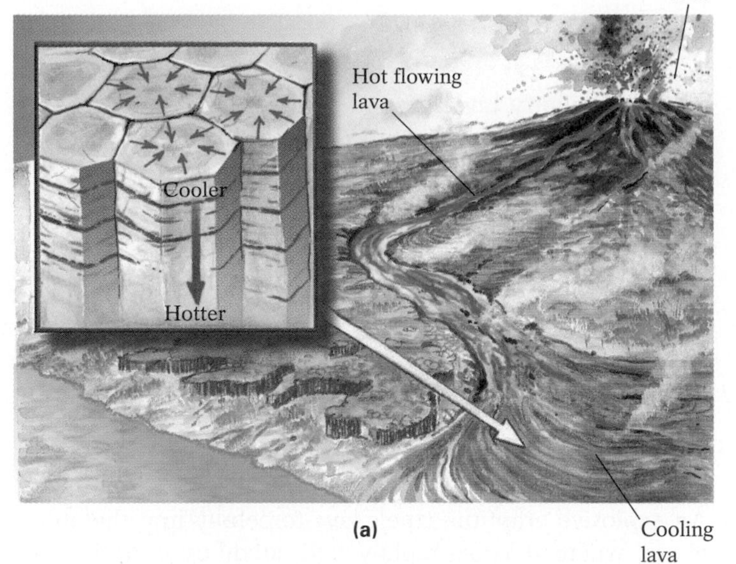

Figure 4-6 Basaltic lava columns. Contraction of basaltic lava flows as they cool **(a)** sometimes produces geometrically patterned columns. **(b)** Such structures can be found in North America in eastern Washington, eastern Oregon, southern Idaho, eastern California, and eastern Wyoming, such as those shown here at Devil's Tower National Monument.

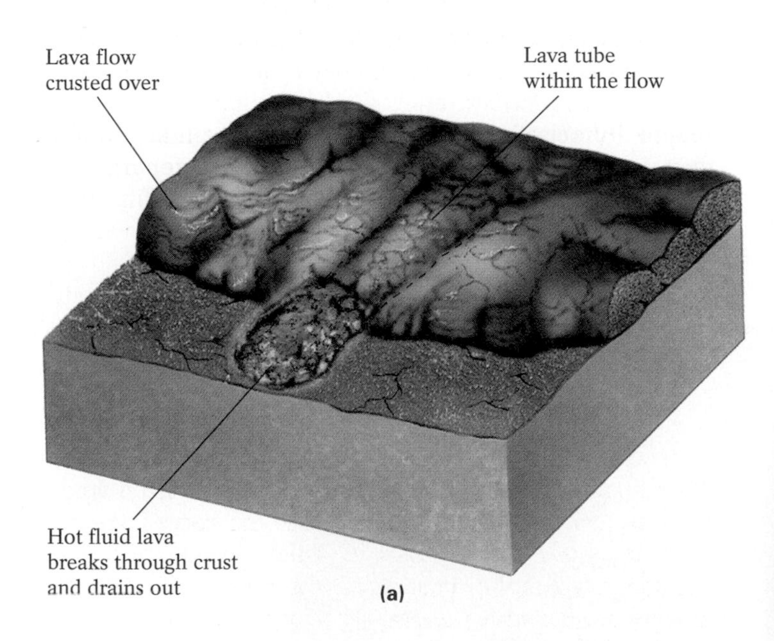

Figure 4-7 Lava tubes. **(a)** Lava tubes form when a lava flow's surface cools and solidifies, but the lava continues to flow under the surface in tunnels. When the lava drains from the tunnel, it leaves behind the empty tube. **(b)** A lava tube forming at K'ilauea.

and left hollow (Fig. 4-7). Lava Beds National Monument in northeastern California contains 300 or more lava tubes in its pahoehoe flows.

When basaltic lava erupts beneath the sea, the contact with cold water instantly chills its surface, which immediately solidifies into a thin, deformable skin. As hot lava enters

Figure 4-8 Basaltic pillow lava in the Galapagos. This form was photographed from a deep-sea submersible.

under it, this skin stretches to form a distinctive *pillow structure* (Fig. 4-8).

Andesitic and Rhyolitic Lavas Andesitic lava, being more felsic and thus more viscous than mafic basaltic lava, flows more slowly than basaltic lava and solidifies before traveling as far from its vent. Such lavas develop many of the same structures as basaltic lava. We rarely see pahoehoe-type andesitic flows, however, because these lavas are too viscous to stretch into a ropy structure. The more felsic andesitic lavas can even be viscous enough to impede the passage of rising gases and erupt in major volcanic explosions.

Felsic magma, being the coolest and most highly viscous form of molten rock, moves so slowly that it tends to cool and solidify underground as plutonic granite. It therefore rarely erupts as rhyolitic lava at the Earth's surface. When rhyolite does erupt, it usually explodes violently, producing an enormous volume of solid airborne fragments. Because it is so viscous, this type of lava never flows far from the vent and does not produce the structures that are typical of less viscous lavas. Felsic magmas with high water and gas content may bubble out of a vent as a froth of lava that quickly solidifies into the glassy volcanic rock known as *pumice*.

Pyroclastics

An explosive eruption expels lava forcefully into the atmosphere, where it cools rapidly and solidifies into countless fragments of various sizes and shapes. Such an eruption might also produce *volcanic blocks,* chunks of preexisting rock ripped from the throat of the volcano during an eruption. Volcanic blocks tend to be angular and range from the size of a baseball to that of a house. Blocks weighing as much as 100 tons have been found as far as 10 kilometers (6 miles) from the volcano that spewed them out. All such fragmental volcanic products, whether they are composed of cooled lava or of preexisting rocks, are termed **pyroclastics** (from the Greek *pyro,* meaning "fire," and *klastos,* meaning "fragments"). Pyroclastic materials may travel through the air as dispersed particles or they may hug the ground as dense flows.

Tephra Pyroclastic particles that cool and solidify from lava that is propelled through the air are called **tephra.** Tephra particles are classified by size, ranging from a fine dust to massive chunks (Fig. 4-9).

(a)

(b)

Figure 4-9 Tephra. Tephra particles range in size from fine dust to large boulders. **(a)** This range can sometimes be found within a single deposit, as here at Mono Craters, California. **(b)** A volcanic bomb, the largest type of tephra.

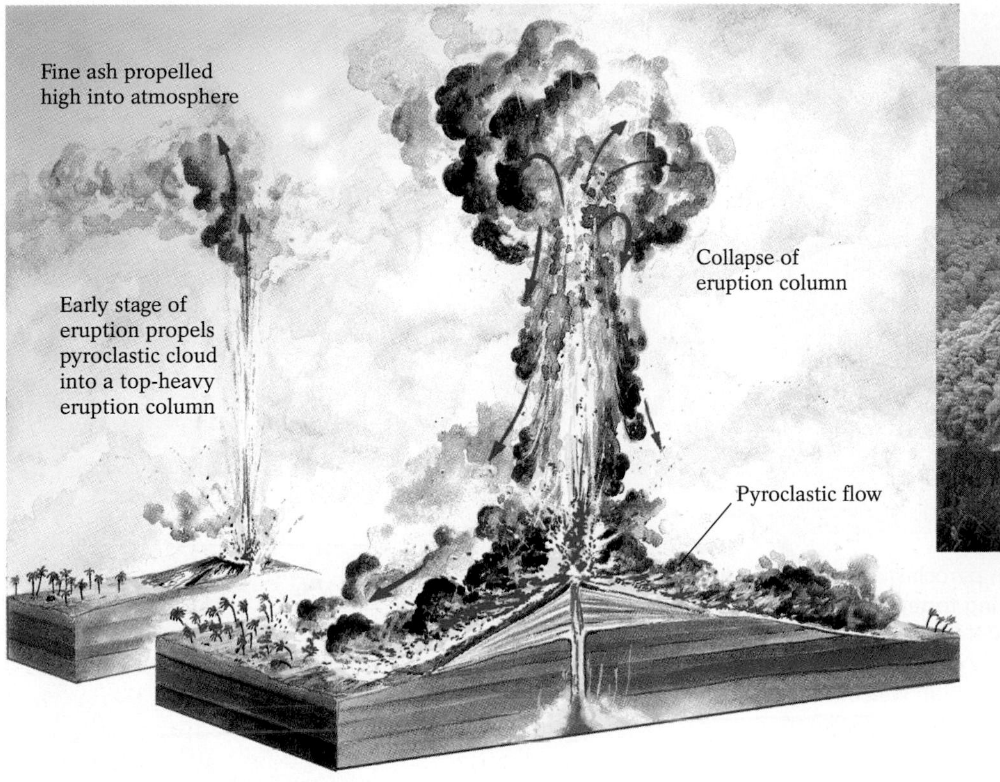

Figure 4-10 Pyroclastic flows. Pyroclastic flows, or nuées ardentes, are produced when a massive amount of airborne pyroclastic material is pulled to Earth by gravity and rushes downslope. Photo: A pyroclastic flow from the May 1991 eruption of Mount Pinatubo in the Philippines.

Volcanic dust particles are only about one-thousandth of a millimeter and have the consistency of cake flour. Because it is so fine, volcanic dust can travel great distances downwind from an erupting volcano and remain in the upper atmosphere for as long as two years. In sufficient quantity, it can diminish the amount of solar radiation reaching the Earth and lower the Earth's temperature by as much as 2° to 3° Celsius, an effect that may last for more than a decade. The spectacular eruption of Indonesia's Mount Tambora in 1815, which took an estimated 50,000 lives and decapitated the volcano's peak, was followed by what became known as the "year without a summer."

Somewhat grittier than volcanic dust is *volcanic ash,* particles of which are less than 2 millimeters in diameter, ranging from the size of a grain of fine sand to that of rice. Ash generally stays in the air for a few hours or days. During the 1980 eruption of Mount St. Helens in Washington state, ash covered the land downwind of the eruption, clogging automobile carburetors and fouling the bearings of farm machinery.

Cinders, or *lapilli* (Italian for "little stones"), range from about the size of peas to that of walnuts (2 to 64 millimeters in diameter). *Volcanic bombs* are large (64 millimeters or more in diameter), streamlined chunks of rock formed when sizable blobs of lava solidify in mid-air while being propelled by the force of an eruption. These coarser types of tephra are pulled to Earth by gravity; hence they fall sooner and closer to the volcanic vent than do dust and ash.

Pyroclastic Flows When the amount of pyroclastic material expelled by a volcano is so great that gravity almost immediately pulls it down onto the volcano slope, this material rushes downslope as a **pyroclastic flow,** or **nuée ardente** (pronounced "noo-AY AR-dent"; French for "glowing cloud"). Because the flow also contains trapped air and magmatic gases, it flows with little frictional resistance along the ground and may reach speeds in excess of 150 kilometers (100 miles) per hour, even on gentle slopes (Fig. 4-10). With temperatures of 800°C (1475°F) or more, a nuée ardente is capable of destroying any life that lies within its path. As the flow travels downslope, its gases escape and the warm particles finally come to rest. At this point, they may still be soft enough to fuse with one another, forming a volcanic rock called *welded tuff.*

Volcanic Mudflows Pyroclastic material that accumulates on the slope of a volcano may become mixed with water to form a volcanic mudflow, or **lahar.** A lahar is often produced when an explosive eruption occurs on a snow-capped volcano, and hot pyroclastic material melts a large volume of snow or glacial ice. One such devastating lahar buried the

Figure 4-11 **Lahars.** Some lahars occur when pyroclastic eruptions melt snow and ice on volcanic slopes, producing torrents of mud. This lahar resulted when Colombia's Nevado del Ruiz volcano erupted in 1985, melting about 10% of the snow on the slopes of the volcano and producing a 40-meter (137-foot)-high wall of mud that buried the town of Armero and killed approximately 23,000 of its residents. Armero was located approximately 50 kilometers (30 miles) from the summit of Nevado del Ruiz.

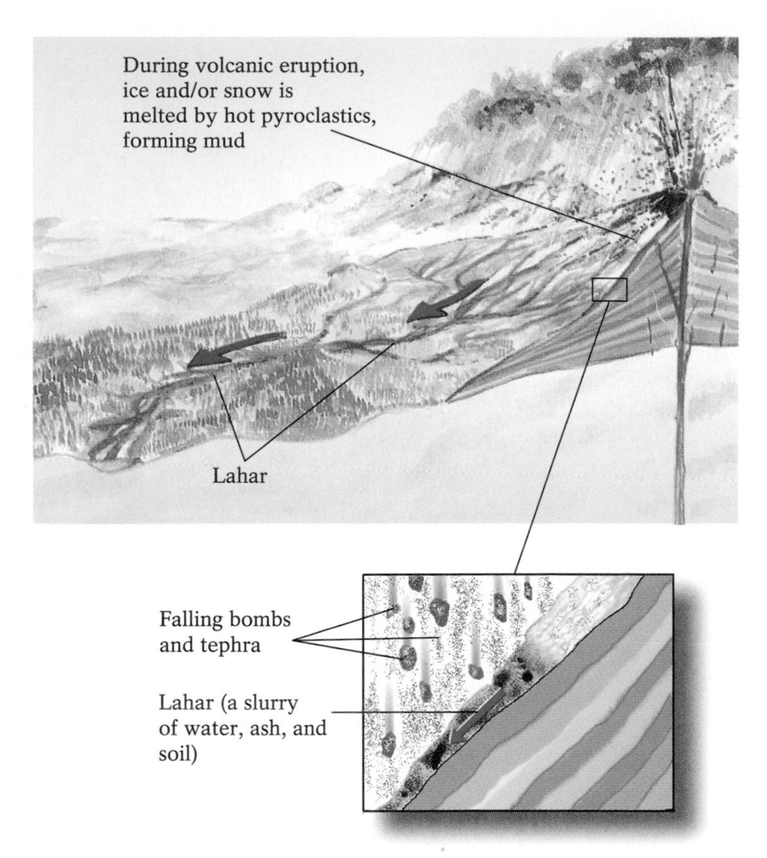

highland town of Armero, on the slopes of Colombia's Andes Mountains, when that country's Nevado del Ruiz erupted on November 13, 1985 (Fig. 4-11).

Eruptive Styles and Associated Landforms

Nearly all volcanoes have the same two major components: (1) a mountain, or **volcanic cone,** constructed by the products of numerous eruptions over time; and (2) a steep-walled, bowl-shaped depression, or **volcanic crater,** surrounding the vent from which those volcanic products emanate. A volcano's crater forms following an eruption, when lava and pyroclastics that have accumulated in the area around the vent are left somewhat unsupported and subside to form a depression. If enough lava erupts to completely or partially empty the volcano's subterranean reservoir of magma, the unsupported summit of the volcanic cone may collapse inward, forming a much larger summit depression called a **caldera.** Volcanic cones and craters take a variety of shapes and dimensions, depending on the types of eruptions and the composition of the volcanic products that formed them (Fig. 4-12).

Effusive Eruptions

Effusive eruptions are relatively quiet, nonexplosive events that generally involve basaltic lava. Basaltic lava, which is highly fluid, flows freely from central volcanic vents, as well as from elongated cracks on land and at submarine plate boundaries.

Central-Vent Eruptions In central-vent eruptions, basaltic lava flows out in all directions from one main vent, solidifying in more or less the same thickness all around. Because of its fluidity, the lava does not develop into a steep mountain. Instead, through successive flows over time, it builds a low, broad, contact lens-shaped structure known as a **shield volcano** (Fig. 4-13). Mauna Loa, one of five such shield volcanoes that form the island of Hawai'i, has a circumference of 600 kilometers (400 miles) and is composed of thousands of layers of lava flows that have erupted over the past 750,000 years.

At the start of an effusive eruption, lava begins to accumulate in the volcanic crater, forming a lava lake that may eventually overflow the rim of the crater. Sometimes the summit collapses to form a caldera, and the weight of the collapsed summit closes off the central vent; in such an event, any remaining magma is diverted laterally, producing a *flank*

(a)

(b)

Figure 4-12 Volcanoes come in various shapes and sizes. **(a)** Lofty, symmetrical Mount Augustine, in Alaska's Cook Inlet. **(b)** Small, stubby volcanoes from the Flagstaff area of northern Arizona.

eruption from the side of the volcano. Flank eruptions also occur when the central vent becomes plugged by congealed lava or when the volcanic cone grows so high that rising magma seeks a lower, more direct route to the surface. Hawai'i's K'ilauea is a volcanic cone built up by flank eruptions on Mauna Loa's southeastern slope.

Fissure Eruptions Rising, highly fluid basaltic lava may erupt through linear fractures, or *fissures*, in the Earth's crust that often develop at diverging plates. As lava flows away from a fissure, it may spread out over thousands of square kilometers; successive flows may build up immensely thick *lava plateaus*, or *flood basalts*, such as the 15-million-year-old

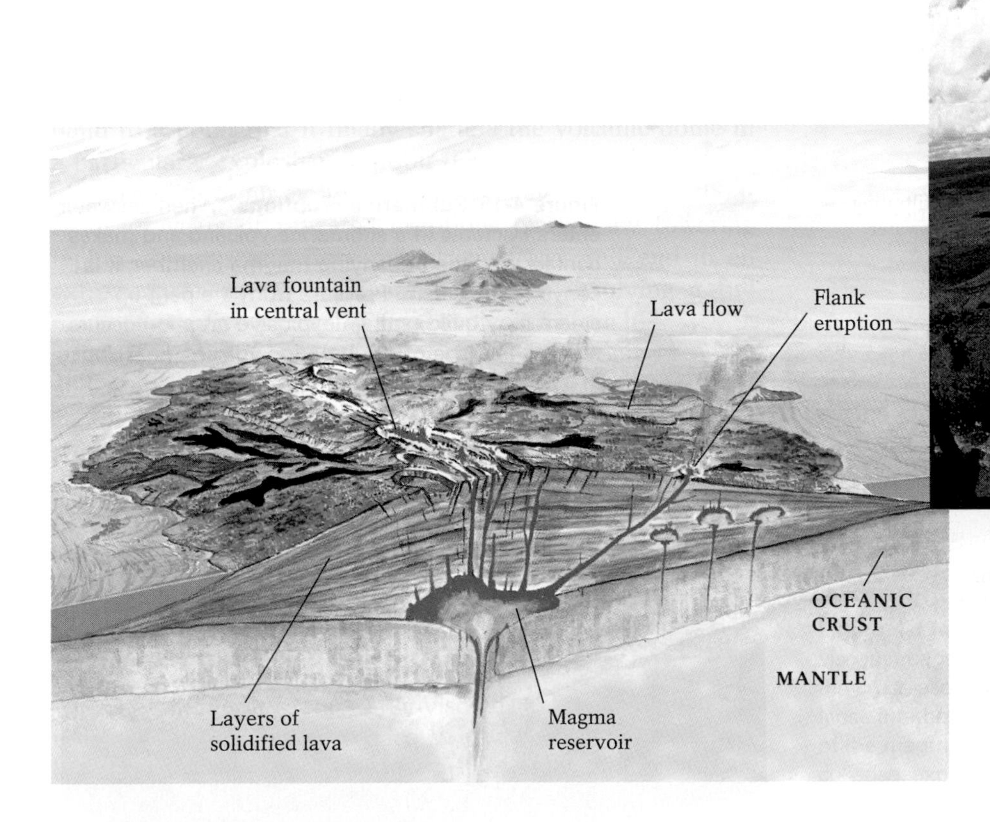

Figure 4-13 Shield volcanoes. Low, broad shield volcanoes form by the gradual accumulation of gently sloping basaltic lava flows. Photo: The summit of Hawai'i's K'ilauea volcano, a shield volcano that actually developed on the flank of an even larger shield volcano, Mauna Loa.

Highlight 4-1 Mount Mazama and the Formation of Crater Lake

The last eruption of Mount Mazama, a volcano in the southern Oregon Cascades, occurred approximately 6700 years ago. The event buried thousands of square kilometers with tephra and pumice and lowered the mountain's height by more than a mile. Before this culminating eruption, the mountain is believed to have been 3700 meters (12,136 feet) high and glacier-covered, comparable in majesty to Mount Rainier or Mount Shasta. Today, it is a sawed-off, goblet-shaped mountain reaching only 1836 meters (6058 feet) above sea level.

Because very little rock from the shattered cone of the volcano has been found in the areas surrounding Mount Mazama—including Oregon, Washington, and British Columbia—geologists have concluded that the summit of the volcano must have collapsed inward during its last eruption, rather than exploding outward, and remains to this day buried beneath the caldera produced by its collapse (Fig. 1). This caldera now contains Crater Lake, North America's deepest lake (more than 600 meters [1900 feet] deep).

The eruption 6700 years ago may not have ended Mount Mazama's excitement, however. Eruptions during the past 1000 years have produced three smaller volcanic cones within the lake, two of which remain below water level. The third, Wizard Island, rises above the surface near the western shore of the lake. Recent dives to the lake bottom reveal that hot water and steam are being vented continuously, suggesting that Mount Mazama may be entering a new eruptive sequence.

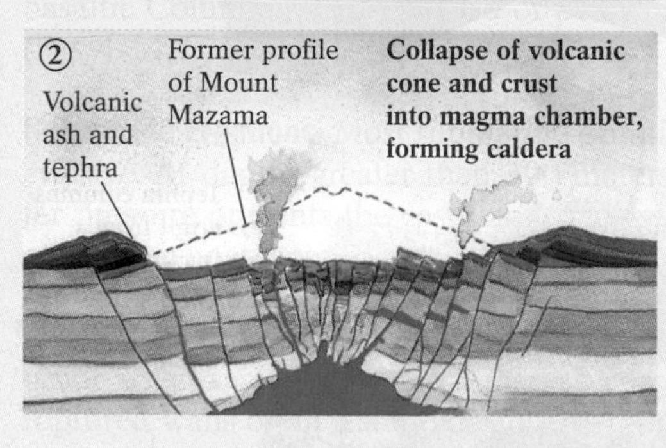

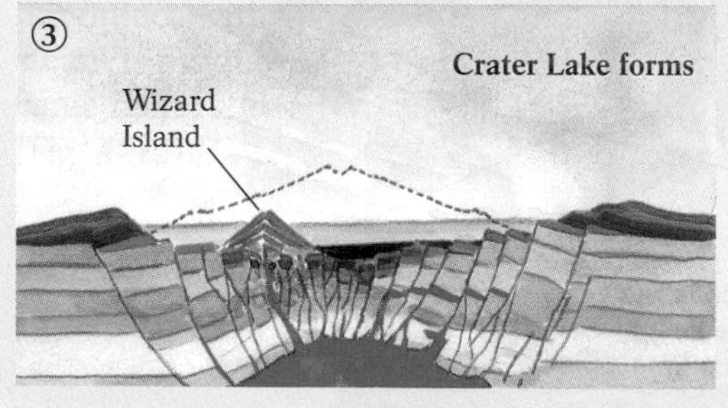

Figure 1 The culminating eruption of Mount Mazama and the creation of Crater Lake.

producing a circular pattern of tremendous tephra columns. These columns soon collapse to form numerous flows of hot, swirling ash and pumice. As the magma reservoir continues to empty, the surface crust becomes further undermined until it collapses to form a huge caldera, sometimes as large as 5000 square kilometers (1800 square miles).

At least three times in the last 2 million years, the Yellowstone plateau in Wyoming has erupted in devastating ash flows after being pushed up like a blister by an enormous mass of felsic magma. One of these events showered thousands of square kilometers with hot ash and pumice and created the caldera that now contains beautiful Yellowstone Lake. Today, only a few kilometers below Yellowstone's caldera, a new huge mass of felsic magma may be accumulating. Yellowstone National Park's thermal features—its geysers, hot springs, and gurgling mudpots (pools of boiling mud)—are all heated by this shallow subterranean magma reservoir.

The most immediate threat of an ash-flow eruption in North America is near Mammoth Lakes, the popular ski resort in eastern California. During the past 20 years, the

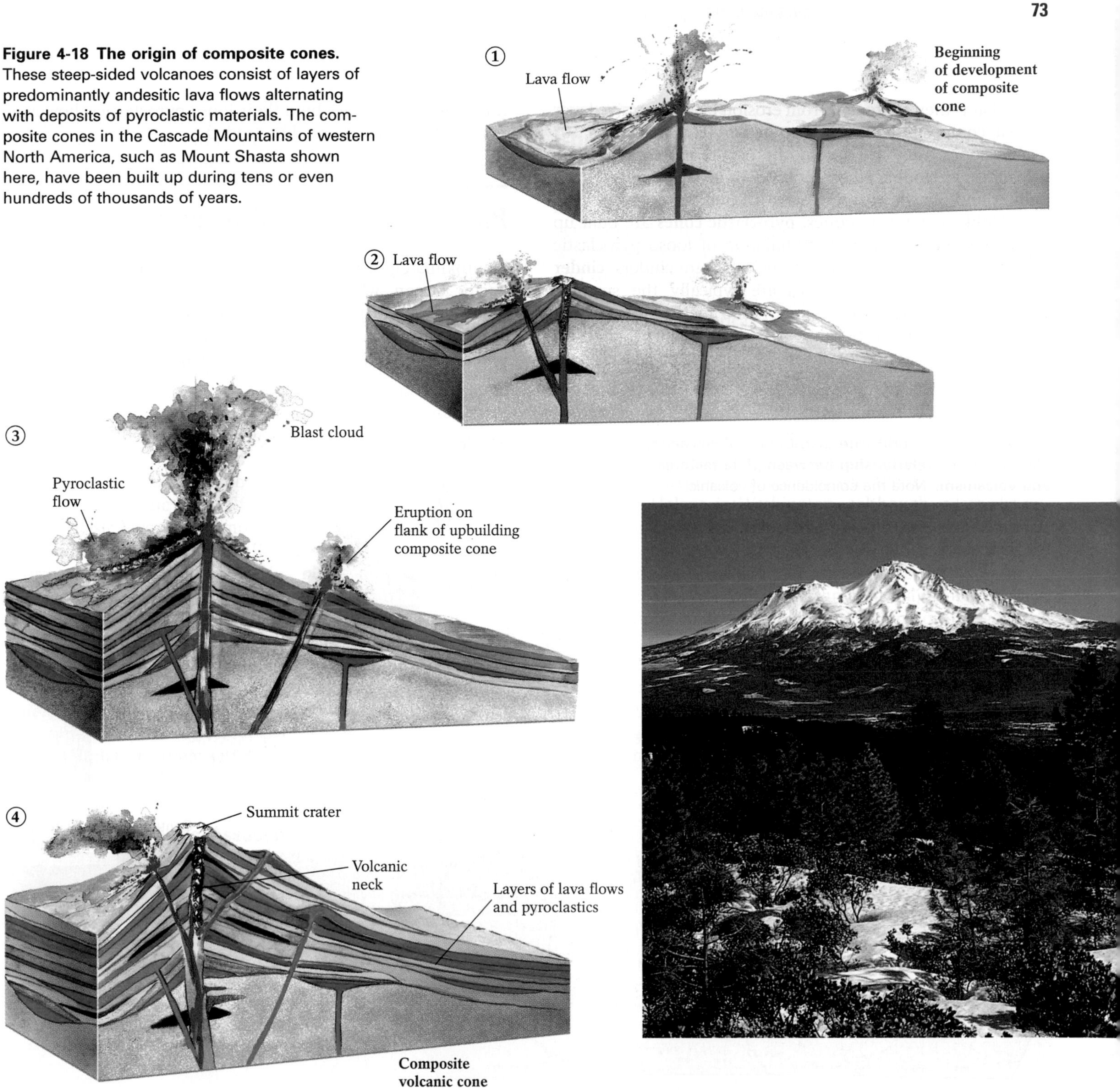

Figure 4-18 The origin of composite cones. These steep-sided volcanoes consist of layers of predominantly andesitic lava flows alternating with deposits of pyroclastic materials. The composite cones in the Cascade Mountains of western North America, such as Mount Shasta shown here, have been built up during tens or even hundreds of thousands of years.

United States Geological Survey has watched the floor of the caldera rise more than 25 centimeters (9 inches). In 1982, it designated the area as a potential volcanic hazard, requiring heightened scientific vigilance and a regional plan for coping with an eruption.

Types of Pyroclastic Volcanic Cones Because they are fairly viscous, felsic and intermediate lavas solidify relatively close to the vent. The composition of the parent magma and thus the volcano's eruptive style may change over time, however, so that the cone intermittently ejects a large quantity of tephra instead of lava. The larger tephra particles from these eruptions fall near the summit to form steep cinder piles, which then become covered by the next lava flow. The characteristic landform of pyroclastic eruptions—the **composite cone,** or **stratovolcano** ("strato" means layered)—is built up from such alternating layers of lava and pyroclastics (Fig. 4-18). Because each pyroclastic deposit produces a steep

Highlight 4-2 Mount St. Helens

Mount St. Helens' cone, the youngest in the Cascades at less than 10,000 years of age, came to life with an audible boom on March 30, 1980, after 123 years of silence. Weeks of public anxiety and scientific watchfulness followed, culminating on May 18 when an eruption blasted about 400 meters (1300 feet) of rock from the summit, changing what many had considered North America's most beautiful peak into a squat gray crater (Fig. 1).

Pre-eruption underground rumblings had signaled the rise of magma into Mount St. Helens' cone. Probably blocked on its way to the central vent by a plug of congealed lava from an earlier eruption, the ascending magma took a sharp turn northward instead. A bulge appeared on the volcano's northern slope and, growing at the rate of 1.5 meters (5 feet) per day, protruded an ominous 122 meters (400 feet) by May 17.

At 8:31 on the morning of May 18, a powerful earthquake beneath the mountain released internal pressure that had been building for weeks. It dislodged the bulge and sent the material hurtling down the mountain as an avalanche of debris traveling at 400 kilometers (250 miles) per hour. A northward-directed jet of superheated (500°C [900°F]) ash and gas immediately erupted as a pyroclastic flow, racing downslope with hurricane force (at speeds greater than 300 kilometers [200 miles] per hour) and cutting a swath of complete destruction 30 kilome-

(a)

(b)

(c)

Figure 1 Before and after the eruption. Mount St. Helens' pre-eruption symmetry made it one of the world's most beautiful and photographed volcanoes **(a)**. Its eruption on May 18, 1980, opened a crater on the volcano's north side **(b)**, changing its appearance dramatically **(c)**.

ters (20 miles) wide. The blast and subsequent nuée ardente buried the nearest 12 kilometers (7 miles) of forest land beneath meters of pyroclastics, blew down entire stands of mature trees like matchsticks to a distance of 20 kilometers (12 miles) (Fig. 2), and singed the forest beyond for an additional 6 kilometers (4 miles). More than 26 kilometers (16 miles) away, the heat scalded fishermen, who plunged into lakes and streams. On Mount Adams, 50 kilometers (30 miles) away, climbers felt a gust of intense heat just before being bombarded with hot, ash-blasted pine cones.

Meanwhile, a tephra column rose from the summit vent to an altitude of 25 kilometers (15 miles). Swept eastward by the prevailing winds, the dense cloud of gray ash began to fall on the cities and towns of eastern Washington. Yakima, located 150 kilometers (100 miles) to the east of Mount St. Helens, received 10 to 15 centimeters (4–6 inches) of ash. In Spokane, farther east, visibility fell to less than 3 meters (10 feet) and automatic street lights switched on at noon. Proceeding across the continent, the ash cloud dusted every state in its path. Hundreds of downwind communities would later spend millions of dollars cleaning up.

Several hours after the eruption began, snow and large chunks of glacial ice trapped within the dislodged debris from the mountain's northern slope began to melt. This enormous volume of meltwater mixed with loose material and the eruption's fresh pyroclastics to produce a lahar that rushed 28 kilometers (17 miles) westward down the Toutle River valley at 80 kilometers (50 miles) per hour, picking up logging trucks and hundreds of thousands of logs along the way. It eventually buried 123 homes beneath 60 meters (200 feet) of mud.

In all, 60 human lives were lost, along with 500 blacktail deer, 200 brown bear, 1500 elk, and countless birds and small mammals. The only survivors were burrowers such as frogs and salamanders, which fled into the soft sands of lake shores and stream banks. The human toll would have been much higher if state officials had not heeded the warnings issued in March by the U.S. Geological Survey, after the volcano's initial reawakening, and evacuated the area of most of its year-round residents and closed it to seasonal residents and spring hikers. However, if the eruption had occurred one day later, on Monday, hundreds of loggers at work would have been buried beneath the debris avalanche.

Figure 2 The awesome power of volcanoes. The forest north of Mount St. Helens was blown down for miles by the force of the volcano's pyroclastic flow and covered with volcanic ash. The area shown in this photo is 12 kilometers (7 miles) from the crater.

Chapter Summary

Volcanism, the set of processes that results in extrusion of molten rock, begins with the creation of magma by the melting of preexisting rock and culminates with the ascent of this magma to the Earth's surface through fractures, faults, and other cracks in the lithosphere. **Volcanoes** are the landforms created when molten rock escapes from vents in the Earth's surface and then solidifies around these vents. Volcanoes may be active, dormant, or extinct.

Because of its high temperature and relatively low silica content, mafic magma has low viscosity (is highly fluid). It generally erupts (as basaltic lava) relatively quietly, or effusively, because its gases can readily escape and do not build up high pressure. Felsic magma, with its high silica content and relatively low temperature, is highly viscous and generally erupts (as rhyolitic lava) explosively.

The nonexplosive volcanic eruptions characteristic of basaltic lava produce lava flows that, when they solidify, are associated with distinctive features such as pahoehoe- and 'a'a-type surface textures, basaltic columns, lava tubes, and pillow structures. The explosive volcanic eruptions characteristic of rhyolitic lavas typically eject **pyroclastic** material—fragments of solidified lava and shattered preexisting rock ejected forcefully into the atmosphere. The various particles produced when lava cools and solidifies as it falls back to the surface are collectively called **tephra.** Explosive ejection of pyroclastic material is usually accompanied by a number of life-threatening effects, such as **pyroclastic flows,** or **nuée ardentes** (high-speed, ground-hugging avalanches of hot pyroclastic material), and **lahars** (volcanic mudflows).

Nearly all volcanoes have the same two major components: (1) a mountain, or **volcanic cone,** built up of the products of successive eruptions; and (2) a bowl-shaped depression, or **volcanic crater,** surrounding the volcano's vent. If enough lava erupts to empty a volcano's subterranean reservoir of magma, the cone's summit may collapse, forming a much larger depression, or **caldera.**

Effusive eruptions, which usually involve basaltic lava, form gently sloping, broad-based cones called **shield volcanoes.** Basaltic magma reaching the surface through long linear cracks, or fissures, in the Earth's crust spreads to produce nearly horizontal lava plateaus.

Explosive **pyroclastic eruptions** involve viscous, usually gas-rich magmas and so tend to produce great amounts of solid volcanic fragments rather than fluid lavas. Felsic (rhyolitic) lavas are often so viscous that they cannot flow out of a volcano's crater; they therefore cool and harden within their craters to form **volcanic domes.** Ash-flow eruptions occur in the absence of a volcanic cone; they are produced when extremely viscous, gas-rich magma rises to just below the surface bedrock, stretching and collapsing it.

The characteristic landform of pyroclastic eruptions is the **composite cone,** or **stratovolcano,** which is composed of alternating layers of pyroclastic deposits and solidified lava. Pyroclastic eruptions may also produce **pyroclastic cones** or **cinder cones,** created almost entirely from the accumulation of loose pyroclastic material around a vent. All pyroclastic-type volcanoes produce steep-sided cones, because the materials they eject—solid fragments and highly viscous lavas—do not flow far from the vent.

Various types of volcanic eruptions are associated with different plate tectonic settings. Explosive pyroclastic eruptions of felsic (rhyolitic) lava generally occur within continental areas characterized by plate rifting or atop intracontinental hot spots. Most intermediate (andesitic) eruptions take place near subducting oceanic plates. Effusive eruptions of (mafic) basalt generally occur at divergent plate margins and above oceanic intraplate hot spots.

Humans can minimize damage from volcanoes by zoning against development in the most hazardous areas, building lava dams, diverting the path of a flowing lava, and learning to predict eruptions accurately. Techniques used to predict eruptions include measuring changes in a volcano's slopes, recording related earthquake activity, and tracking changes in the volcano's external heat flow.

Volcanism is not restricted to the Earth. It has occurred in the past on the Moon, and relatively recent volcanic activity has been detected on Mars, Venus, and the moons of Jupiter and Neptune.

Key Terms

volcanoes (p. 61)
volcanism (p. 62)
pyroclastics (p. 66)
tephra (p. 66)
pyroclastic flow (p. 67)
nuée ardente (p. 67)
lahar (p. 67)
volcanic cone (p. 68)
volcanic crater (p. 68)
caldera (p. 68)
shield volcano (p. 68)
pyroclastic eruptions (p. 70)
volcanic dome (p. 71)
composite cone (p. 73)
stratovolcano (p. 73)
pyroclastic cones (p. 74)
cinder cones (p. 74)

Questions for Review

1. What criteria do geologists use to designate a volcano as active, dormant, or extinct?

2. Briefly compare basaltic, andesitic, and rhyolitic lava, in terms of their composition, viscosity, temperature, and eruptive behavior.

3. Within a single basaltic lava flow issuing from a Hawai'ian volcano, why is pahoehoe lava found closer to the vent than 'a'a lava?

4. Contrast the nature and origin of nuée ardentes and lahars.

5. Describe the three basic types of effusive eruptions (central-vent, fissure, submarine) and the volcanic landforms associated with each.

6. How does a composite cone form? What type of lava is associated with a composite cone? How could you distinguish a composite cone from a pyroclastic cone?

7. What types of volcanoes and volcanic landforms are associated with subduction zones? With divergent plate boundaries?

8. Identify three sites in North America that pose a volcanic threat to nearby residents.

9. Describe three techniques that geologists use to predict volcanic eruptions.

10. Compare the volcanism on the Moon to volcanism on Io, Jupiter's moon.

For Further Thought

1. Look at the photograph below and speculate about the plate tectonic setting where this volcano is found; the composition of the rocks that make up the volcano; and whether eruptions of this volcano tend to be explosive or effusive.

2. The 1980 eruption of Mount St. Helens made a lot of headlines but had no discernible effect on global climate. Conversely, the eruption of the Philippines' Mount Pinatubo in 1991 caused a 1°C drop in global temperature. What differences between these eruptions might explain why one had a sharp effect on climate, whereas the other did not?

3. Why do we find andesite throughout the islands of Japan, but not throughout the islands of Hawai'i?

4. How would you explain the origin of a volcanic structure composed of 10,000 meters of pillow lava covered by 3000 meters of basalt containing vesicles, basaltic columns, and 'a'a and pahoehoe structures?

5. Under what circumstances might active volcanism resume along the east coast of North America? Within the Great Lakes region of North America?

5

Weathering: The Breakdown of Rocks

Much of the Earth's most spectacular and unique scenery has been created by quite ordinary environmental factors acting upon rock. These factors, which act constantly and everywhere on Earth, remove or alter individual mineral grains in rock, yielding end products that look much different from the original rocks. The processes by which environmental agents at or near the Earth's surface cause rocks and minerals to break down is called **weathering.** Weathering is a slow but potent force to which even the hardest rocks are susceptible (Fig. 5-1).

Rocks that have been weakened by weathering are more vulnerable to **erosion,** the process by which gravity, moving water, wind, or ice transports pieces of rock and deposits them elsewhere. The fragments removed from rock by erosion have usually been loosened by weathering, although a high-energy erosive event, such as a flood, may dislodge even unweathered rock. Like many geological processes, weathering and erosion are interrelated and often work in tandem. Together, they produce *sediment,* the loose, fragmented surface material that serves as the raw material for sedimentary rock (discussed in Chapter 6).

Weathering plays a vital role in our daily lives, with both positive and negative outcomes. It frees life-sustaining minerals and elements from solid rock, allowing them to become incorporated into our soils and finally into our foods. Indeed, we would have very little food without weathering, as this process produces the very soil in which much of our food is grown. But weathering can also wreak havoc on the structures we build. Countless monuments—from the pyramids of Egypt to ordinary tombstones—have suffered drastic deterioration from freezing water, hot sunshine, and other climatic forces.

Figure 5-1 The beauty of weathering. Monument Valley in northeastern Arizona shows the results of millions of years of exposure to the Earth's environment. The rocks that were removed to produce these landforms were less resistant to weathering and erosion than are the hardy rocks that remain.

Weathering Processes

Rocks can be weathered in two ways. **Mechanical weathering** breaks a mineral or rock into smaller pieces (*disintegrates* it) but does not change its chemical makeup.

6

The Origins of Sedimentary Rocks

Classifying Sedimentary Rocks

Highlight 6-1
When the Mediterranean Sea Was a Desert

"Reading" Sedimentary Rocks

Sedimentation and Sedimentary Rocks

Figure 6-1 **Layers of sedimentary rock (Navajo Sandstone) in Zion National Park, Utah.** These rocks are composed of the cemented sand grains of ancient sand dunes. The fascinating "cross-bedded" patterns they exhibit are typical of windblown sands.

Sediment, the unconsolidated material that accumulates continuously at the Earth's surface, consists mostly of the physical and chemical products of weathering and erosion discussed in Chapter 5: solid fragments of preexisting rocks, and minerals precipitated out of solution in water. The organic remains of plants and animals also contribute to sediment. After being transported by water, wind, or ice or precipitating out of solution, sediment accumulates virtually everywhere on Earth—from the glaciated summits of the Himalayas, 10 kilometers (6 miles) above sea level, to the deep trenches on the floor of the Pacific Ocean, 10 kilometers below sea level. It is continually deposited throughout the world—in lakes, streams, oceans, swamps, beaches, lagoons, deserts, caves, and at the bases of glaciers.

Much sediment is ultimately converted to solid **sedimentary rock** (Fig. 6-1). Sedimentary rocks make up only a thin layer of the Earth's uppermost crust, accounting for barely 5% of the Earth's outer 15 kilometers (10 miles), but they constitute 75% of all rocks exposed at the Earth's land surface. Sedimentary rocks serve as our principal source of coal, oil, and natural gas, as well as much of our iron and aluminum ores; they also store nearly all of our fresh underground water and represent the source of cement and other natural building materials.

In addition, sedimentary rocks offer clues about the condition of the Earth's surface as it existed in the geologic past. They record the former presence of great mountains in areas now monotonously flat, and they tell tales of vast seas that once covered the now dry interior of North America. Some sedimentary rocks contain the fossil remains of past life, revealing much about how Earth has evolved through its history.

This chapter examines the origins of sedimentary rocks, describes their classification, and explains the ways in which geologists use them to reconstruct past surface environments. It concludes by showing how various sedimentary rocks relate to common plate tectonic settings.

Conditions Promoting Metamorphism

Types of Metamorphism

Common Metamorphic Rocks

Metamorphic Grade and Index Minerals

Metamorphic Facies and Plate Tectonics

Metamorphic Rocks in Daily Life

Looking Ahead

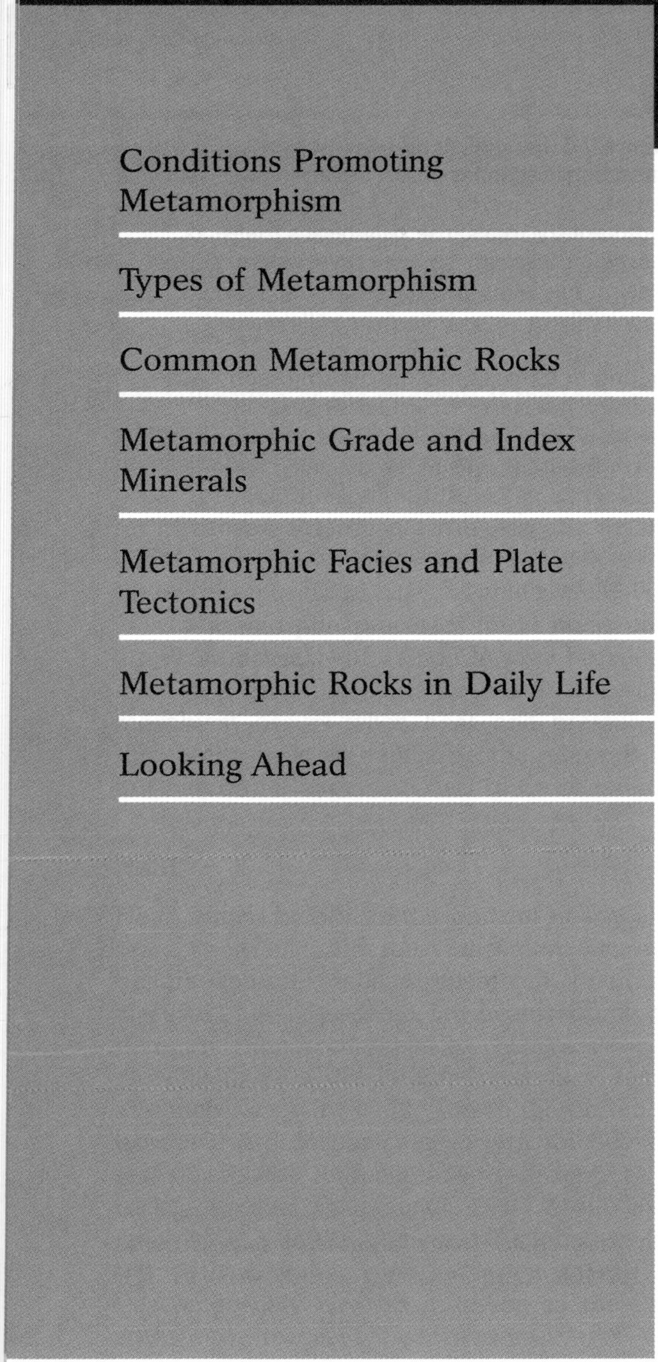

7

Metamorphism and Metamorphic Rocks

We saw in Chapter 3 that high temperatures deep in the Earth's interior create the magmas that eventually cool to form igneous rocks. We saw in Chapter 6 that sediments lithify to become sedimentary rocks in the relatively low-temperature environment near the Earth's surface. The third major type of rock, **metamorphic rocks,** generally forms at conditions between those that produce igneous and sedimentary rocks (Figs. 7-1 and 7-2).

Metamorphism is the process by which temperature, pressure, and chemical reactions deep within the Earth—but above the 50- to 250-kilometer–deep "melting zone" that creates most magmas—alter the mineral content and/or

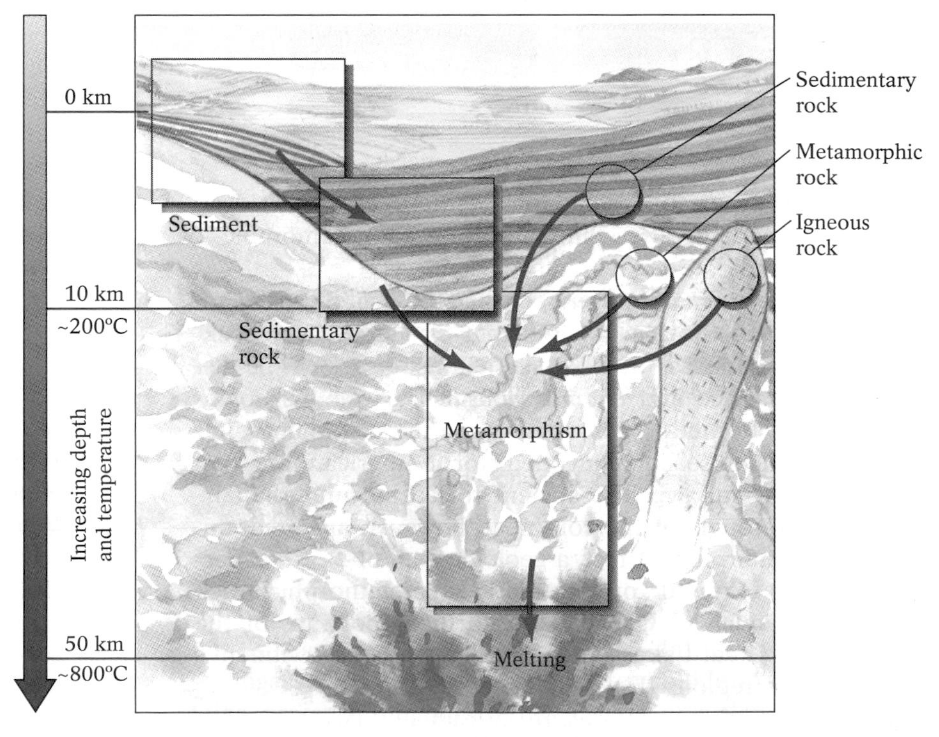

Figure 7-2 Sedimentary, metamorphic, and igneous rocks. Any type of rock may be metamorphosed at temperatures and pressures greater than those that lithify sediment into sedimentary rock, but less than those that melt rock into magma.

Figure 7-1 The beauty of metamorphic rocks. These picturesque rocks at Pemaquid Point in coastal Maine have been converted to metamorphic rocks by extreme heat and pressure.

9

Folds, Faults, and Mountains

In the rocks along Sagami Bay near Yokohama, Japan, lives a colony of clams called *Lithophaga*, or "rock eaters." These creatures scoop out small shelters for themselves from the soft rocks at sea level and wait there for high tide to flood their homes, bringing them meals of marine algae. Moments after Japan's great earthquake of 1923, the land at Sagami Bay shifted upward, leaving rows of *Lithophaga* 5 meters (16 feet) above sea level. Several rows of abandoned *Lithophaga* dwellings are found even higher, in the cliffs at Sagami Bay—one that correlates with the area's 1703 tremor and another that correlates with an earthquake occurring in 818. The rocks adjacent to the bay have risen roughly 15 meters (50 feet) during the last 2000 years. At Sagami Bay we are witnessing the building of a mountain.

The Rockies, the Alps, the Andes, and the Himalayas have all been shaped from common rocks during the course of millions of years. In the Pacific Northwest, the Cascade Mountains continue to rise higher even as you read this book. In locales all over the world, we can see rocks that have been twisted and bent or that have broken and shifted position in less dramatic fashion (Fig. 9-1). What enormous forces could distort solid rock in these ways? With few exceptions, all such features owe their existence to plate tectonics. The same forces that carry the Earth's plates can tear the edges of those plates apart, rupturing them into huge displaced blocks, or squeeze plate edges together, crumpling and uplifting them into great folds of rock.

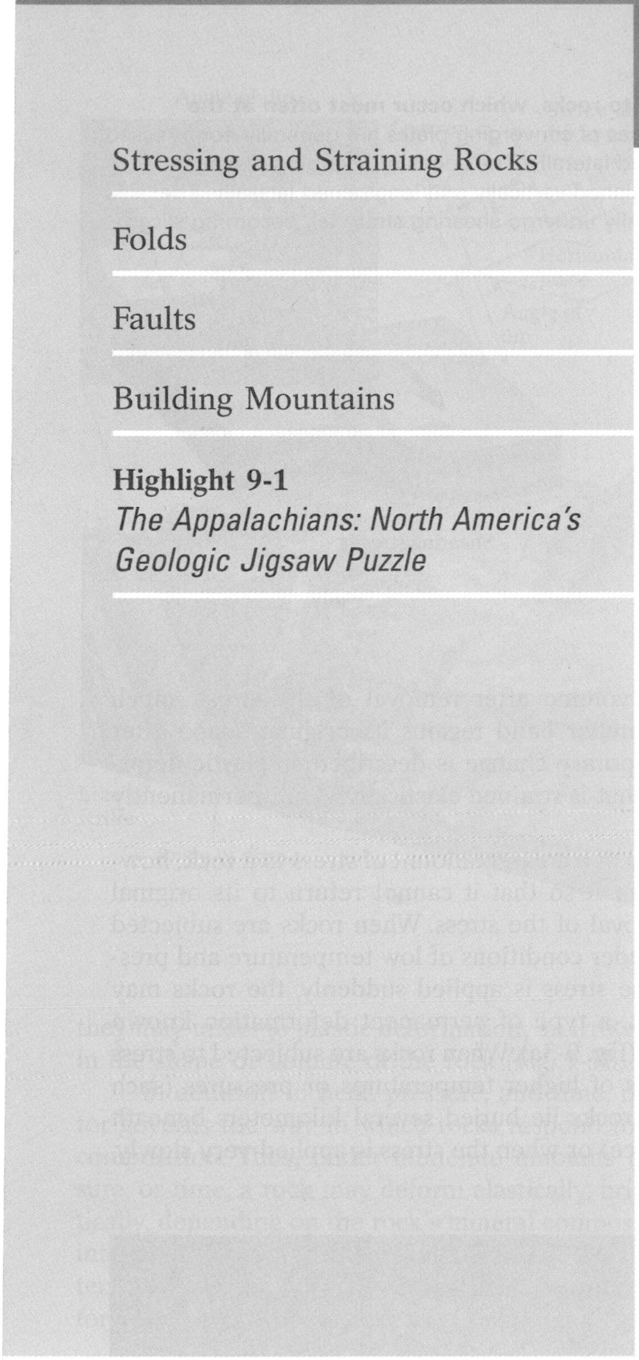

Figure 9-1 Rock distortion. These rocks in the Lower Ugab valley, Namibia, have been tilted and bent by powerful plate-tectonic forces.

Stressing and Straining Rocks

When tectonic plates interact, crustal rocks near the plate boundaries are subjected to a powerful force, or **stress,** that deforms them, changing their shape and often their volume. Rocks may be stressed in three ways, corresponding to three basic types of plate-boundary movements. Rocks at converging plate margins are pushed together, a type of stress

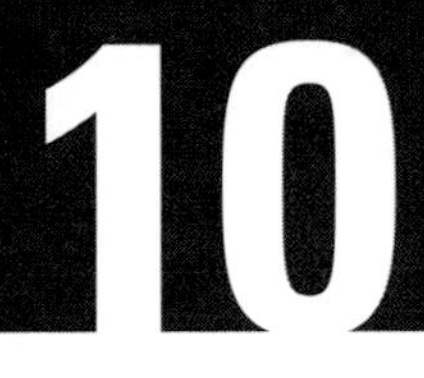

Earthquakes and the Earth's Interior

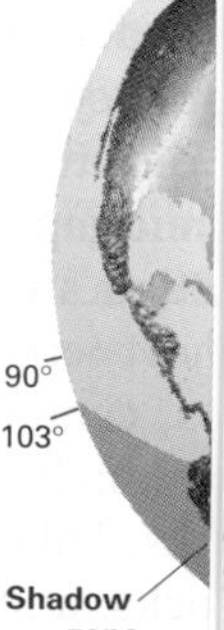

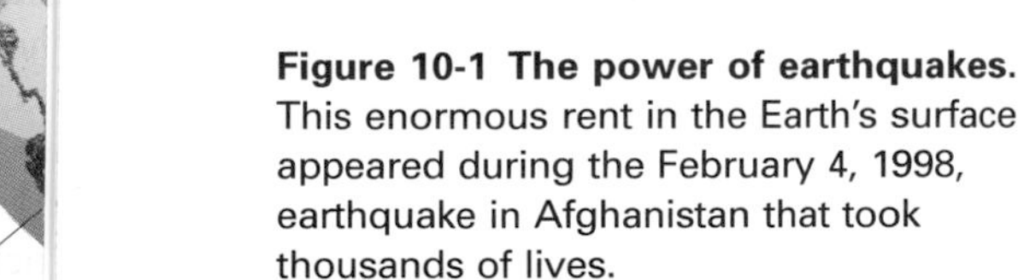

Figure 10-1 The power of earthquakes. This enormous rent in the Earth's surface appeared during the February 4, 1998, earthquake in Afghanistan that took thousands of lives.

At 5:46 A.M. on January 17, 1995, a powerful earthquake, calculated at 7.2 on the Richter scale, hit western Japan. The earthquake, which was centered near the port city of Kobe on Awajishima Island, left more than 5000 dead and at least 29,000 injured. Approximately 30,000 residences in and around Kobe were destroyed or severely damaged, leaving 310,000 people homeless. Hanshin Expressway, connecting Kobe and Osaka, 31 kilometers (19 miles) away, collapsed in five places. Rail transportation, including the Shinkansen bullet train service, was disrupted for hundreds of kilometers, and at least seven derailments were reported. The cost of cleanup and restoration in the aftermath of the Kobe earthquake was expected to reach as much as $150 billion.

In December 1995, an even more powerful earthquake occurred 5 kilometers (3 miles) off the west coast of Mexico. The quake, which struck some of Mexico's most popular beach resorts, toppled houses and hotels, cracked bridges, opened meter-wide fissures in the main coastal highway, and cut power and phone service throughout the region. Damage was greatest in the resort town of Manzanillo, where the luxury seven-story Costa Real Hotel collapsed, killing at least 20 guests and staff. Even in Mexico City, located 335 kilometers (201 miles) to the east, earthquake vibrations caused skyscrapers to sway violently.

In the United States, two major earthquakes have occurred relatively recently. The powerful Loma Prieta earthquake struck northern California in 1989, toppling buildings, freeways, and bridges, rupturing gas mains and starting fires, and setting off landslides throughout the San Francisco Bay area. In 1994, another earthquake produced severe damage in the Northridge area of southern California. Together, these two quakes resulted in billions of dollars of damage and caused more than 100 deaths.

In this chapter, we will discuss what causes earthquakes (Fig. 10-1), how geologists study, evaluate, and even predict them, and how the study of earthquake waves allows us to investigate the Earth's interior.

between t
mantle—l
the **Moho**

The Man

The mantl
of the Ea
the planet
meters (18
silicates th
pyroxene.
steadily as
at certain
seismic-wa
eralogical
sublayers
zone, and

The Upper
compositio
metamorpl
calcium-ric
morphic m
The increa
depth, how
teristics of

Direc
tle, rocky
miles) bene
ing P wave
ble crust ir
the upper r
sphere. Ber
to the poin
tially molte
In this asthe
are capable
creases wh
the asthenc
meters (44
mainder of
structurally
therefore th
again in thi

The Lower
2900 kilom
Despite the
gion, pressu
per mantle
This pressu
tures and, w
in the densi

11

Plate Tectonics: Creating Oceans and Continents

Figure 11-1 The Himalayas in Tibet. Located at the boundary between India and China, this range is a spectacular example of the mountain-building capacity of colliding lithospheric plates.

Virtually every aspect of geology—from the origin of earthquakes, volcanoes, and mountains (Fig. 11-1) to the rise and fall of sea level—is affected by plate motion. In Chapter 1, we made our initial examination of the theory of plate tectonics. In later chapters, we detailed several of the basic assumptions of this theory:

- The Earth's lithosphere consists of rigid plates averaging 100 kilometers (60 miles) in thickness, ranging from about 70 kilometers (43 miles) thick for the oceans to 150 kilometers (90 miles) thick for the continents.
- The plates move relative to one another by divergence, convergence, or transform motion.
- Oceanic lithosphere forms at divergent plate boundaries and is consumed at subduction zones, one type of convergent plate boundary.
- Three basic types of convergent plate boundaries exist: those that occur where ocean plates subduct beneath other ocean plates; those that occur where ocean plates subduct beneath adjacent continents; and those that occur where two continental plates collide.
- Most earthquake activity, volcanism, faulting, and mountain building takes place at plate boundaries.
- Plates generally do not deform internally; that is, the centers of plates tend to be geologically stable.

In this chapter, we take a large-scale view of plate motion and its effects. We examine the rates of plate motion, the origin of both the oceanic and continental portions of the plates, and the driving mechanisms for plate motion.

Key Terms

continental drift (p. 196)
hot spots (p. 199)
marine magnetic anomaly (p. 200)
passive continental margins (p. 203)
ophiolite suite (p. 204)
mid-ocean ridge (p. 204)
ocean trenches (p. 206)
volcanic arc (p. 207)
suture zone (p. 207)
continental shields (p. 209)
continental platform (p. 210)
craton (p. 210)
displaced terranes (p. 211)
microcontinents (p. 211)
thermal plumes (p. 213)

Questions for Review

1. How could you use the distribution of modern and ancient animals to show that continents drift?

2. Why did the scientific community initially reject Wegener's ideas of continental drift? Why were his ideas finally accepted?

3. Describe three methods for determining the velocity of plate motion.

4. What is an aulacogen? How does it form?

5. Describe what happens at a rift margin, from the onset of rifting to the formation of an ocean.

6. Draw a simple sketch of an ophiolite suite (include all of the rock types and structures).

7. Explain why the East Pacific rise, although it is a mid-ocean ridge, is not in the center of the Pacific Ocean.

8. Sketch a convergent boundary between an oceanic plate and a continental plate. Be sure to include all of the important associated landforms.

9. Describe the basic components of a continent. Why are a continent's oldest rocks always in its interior?

10. Briefly describe the fundamental differences between the convection-cell hypothesis and the thermal-plume hypothesis of plate tectonics.

For Further Thought

1. When geologists find ancient glacial deposits in equatorial Africa, they usually interpret them as polar deposits that have drifted from a cold place to a warm place. Formulate another hypothesis to explain this phenomenon.

2. Of the two patterns of marine magnetic anomalies shown below, which shows a faster rate of sea-floor spreading? Explain.

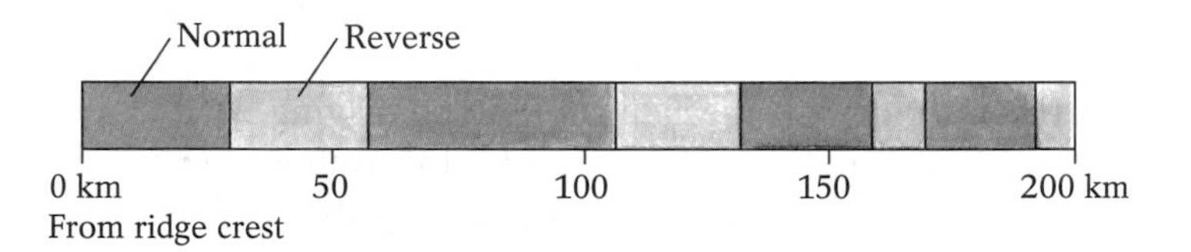

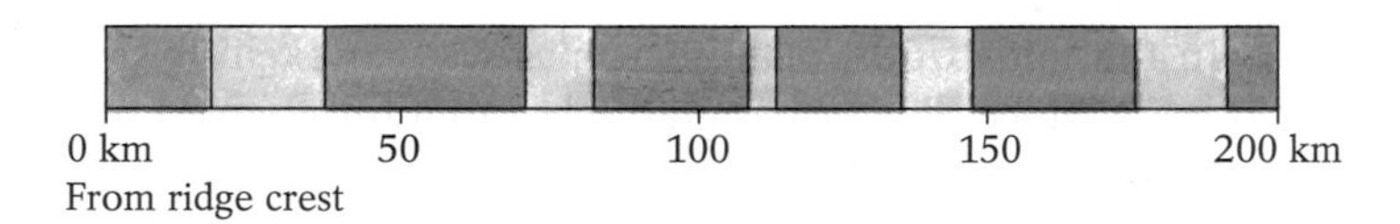

3. What would be some of the major geologic repercussions if a new rift opened between Ohio and Indiana?

4. How would the eastern coast of North America change if the oceanic segments of the plates that make up the Atlantic Ocean basin began to subduct? How would it change if the Atlantic Ocean lithosphere subducted completely?

5. How might the development of a new subduction zone along the East Coast affect plate interactions on the West Coast?

Part 3 Sculpting the Earth's Surface

12

Mass Movement

At 11:37 P.M. on August 17, 1959, an earthquake registering 7.1 on the Richter scale shook thousands of square kilometers in the vicinity of West Yellowstone, Montana. The surface of the ground rolled like sea waves, and the water of lakes and rivers sloshed back and forth. A wall of water rushed across Hebgen Lake and over its dam, sweeping away campgrounds on its way downstream. The quake also triggered a landslide involving more than 80 million tons of rock and weathered regolith, including boulders as large as 9 meters (30 feet) in diameter, as material became dislodged from a steep canyon wall and hurtled downslope at speeds reaching 150 kilometers (90 miles) per hour. As the slide tumbled into Madison Canyon, it compressed and forcefully expelled the air in its path, producing hurricane-force winds that battered the valley. Two-ton cars were blown into the air; one flew more than 10 meters (33 feet) before being dashed against a tree. By the time the slide mass finally came to rest, it had covered the valley floor with 45 meters (150 feet) of bouldery rubble, buried U.S. Highway 287, and taken the lives of 28 campers.

The Madison Canyon disaster is an extreme example of **mass movement,** the process that transports quantities of Earth materials (such as bedrock, loose sediment, and soil) down slopes by the pull of gravity (Fig. 11-1). *Every* slope is susceptible to mass movement. Sometimes the movement is as fast as the Madison Canyon landslide and involves a great mass of material that may travel more than 80 kilometers (50 miles) from its source. More often, the movement is so slow as to be imperceptible and involves just the upper few centimeters of loose soil on a gentle hillside.

Of the 20,000 lives lost as a result of all natural disasters in the United States during the years 1925 to 1975, fewer than 1000 deaths were a direct result of mass movement. Damage associated with mass movement, however, cost a staggering $75 billion, compared with the $20 billion in damage associated with all other natural catastrophes. Thus, although very few of us are likely to perish in a mass-movement event, a strong likelihood exists that we will incur some costs—perhaps because of a cracked house foundation or a living room filled with flowing mud—as a result of this geological process.

Figure 12-1 Mass movement. This massive rockfall occurred in January 1997 along Highway 140, several kilometers east of the Arch Rock entrance to Yosemite National Park, California.

In this chapter, we examine both the underlying and immediate causes of mass movement, the various types of mass movements, and some of the ways their dangers can be prevented.

What Causes Mass Movement?

The principal factor driving all mass movement is gravity, which constantly coaxes materials downslope. Two main factors provide resistance to mass movement: the friction between a slope and the loose material at its surface, and the strength and cohesiveness of the material composing the slope, which prevents it from breaking apart and slipping at its surface. Mass movement occurs when gravity overcomes these factors. The steepness of the slope, the water content of its materials, the amount of vegetative cover, and the slope's history of human and other animal disturbances all influence the mass-movement potential of a slope.

Steepness of Slope

The pull of gravity downslope is proportional to the steepness of the slope. Thus, the steeper a slope, the more likely that material will slide down it. Several natural and artificial processes can steepen slopes and initiate mass movements. For instance, the natural forces of rivers and coastal waves can steepen slopes by undermining them at their base; human activities that can steepen and destabilize slopes include quarrying, road cutting, and dumping excessive amounts of mining waste (Fig. 12-2). Faulting, folding, and tilting of strata can also steepen and destabilize slopes.

Slope Composition

A slope may be composed of any combination of solid bedrock, weathered bedrock, and soil, along with varying quantities of vegetation and water. A slope of solid bedrock tends to be highly stable, even when it is so steep that it forms a vertical cliff. This stability can be greatly reduced, however, if tectonic stresses or weathering (such as frost wedging, root penetration, or chemical dissolution) create cracks or cavities in the rock. Slopes are also less stable when they are parallel to bedding planes or cleavage planes in the rock (Fig. 12-3).

A slope composed of loose, dry particles remains stable only if the friction between its components exceeds the downslope pull of gravity, which increases with the steepness of the slope. Different materials form stable slopes at different angles, depending on the size, shape, and arrangement of their particles. In general, large, flat, angular grains with rough, textured surfaces and a chaotic arrangement create more friction and can form steeper slopes than can smaller grains that are rounded, smooth, and deposited in parallel planes. The maximum angle at which an unconsolidated material can form a stable slope is known as its **angle of repose.** Dry sand, for example, has an angle of repose ranging from 30° to 35°; once a pile of dry sand has attained this slope,

Figure 12-2 Some common processes that oversteepen slopes.

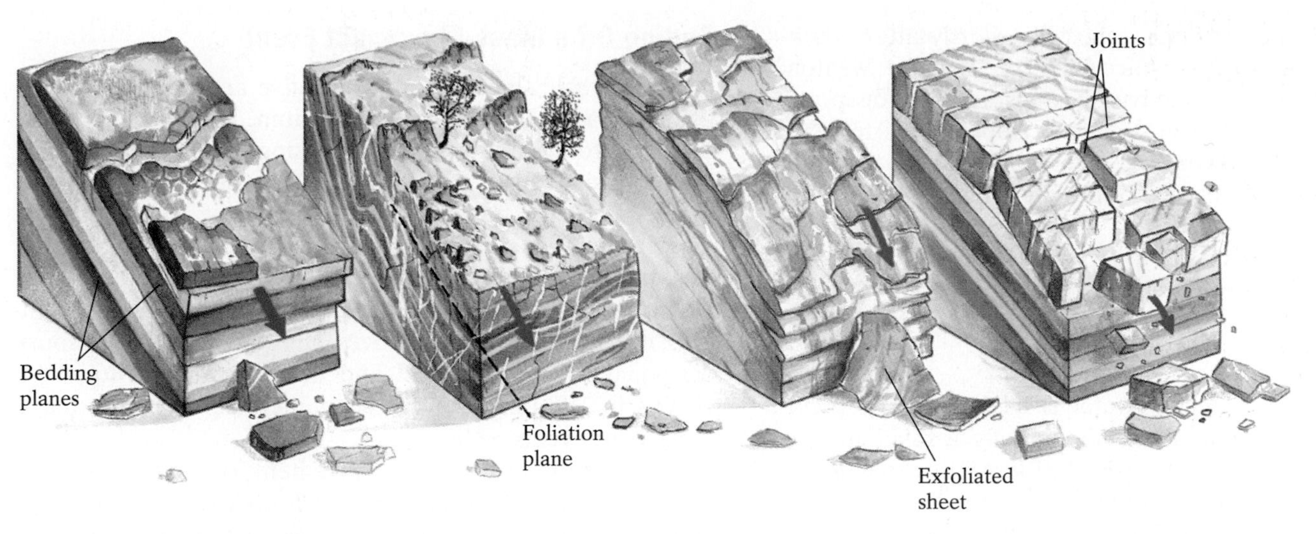

Figure 12-3 Mass-movement potential. Slopes such as these, with planes of weakness parallel to their surface, are especially susceptible to mass movement.

any extra sand added to it would simply cascade down its sides and accumulate at the base of the pile (Fig. 12-4a). In nature, the steepest slopes, greater than 40°, are maintained by large, highly angular boulders. These large boulder piles, called *talus slopes,* often form at the feet of cliffs that have been weathered by frost wedging (Fig. 12-4b).

Vegetation

Vegetation—especially the extensive and deep root networks of large shrubs and trees—binds and stabilizes loose, unconsolidated material. Removal of vegetation by forest fire or clear-cutting for timber or farming purposes allows loose

Figure 12-4 The relationship between angle of repose for unconsolidated materials and particle size and shape. (a) Coarse, angular sand forms a steeper slope than does fine, rounded sand. **(b)** A talus slope, composed of large and irregularly shaped boulders, can form slopes greater than 40°. Photo: Talus slope at Wheeler Peak, Great Basin National Park, Nevada.

material to move downslope, especially shortly after a rainstorm. Several decades ago, farmers in the town of Menton, France, decided to remove olive trees—which have deep, stabilizing roots—from the area's steep slopes so as to plant more profitable shallow-rooted carnations. This horticultural miscalculation contributed to landslides that took 11 lives. Similar widespread tree-cutting in the forests of the Philippines was a primary cause of the tragic landslides that claimed 3400 lives in November 1991.

Water

More than any other factor, water is likely to cause previously stable slopes to fail and slide. Initially, a small amount of water increases the cohesiveness of loose material; damp sand, for example, holds together more effectively than dry sand. Some water also enables the growth of stabilizing vegetation. Excessive water, however, can promote slope failure by reducing friction. It can diminish the friction between surface materials and underlying rocks, between adjacent grains of unconsolidated sediment (for example, turning relatively stable soil into flowing mud), or even between adjacent rock masses that are separated by a plane of weakness (such as a bedding plane, fault, or joint) (Fig. 12-5). Water also promotes slope failure by adding to the weight of slope materials, making them more susceptible to the pull of gravity.

Setting Off a Mass-Movement Event

Before a stable slope becomes unstable and fails, it may develop a fragile balance, or equilibrium, between the forces that tend to drive movement downslope and the forces that tend to resist movement. Some event, either natural or man-made, may then tip the balance, triggering the downslope movement. Natural triggers have included torrential rains, earthquakes, and volcanic eruptions. In 1967, a three-hour downpour in central Brazil triggered hundreds of slope failures, taking more than 1700 lives. The New Madrid, Missouri, earthquakes of 1811–1812 and the eruption of Mount St. Helens in 1980 each produced numerous, damaging mass-movement events.

Human-induced mass movement often results from mismanagement of water. When we overirrigate slopes for farming, install septic fields that leak sewage, divert surface water onto sensitive slopes at construction sites, or over-water sloping lawns, we introduce liquids that destabilize slopes by reducing friction. In one case of inadvertent water mismanagement, a Los Angeles family went on vacation and left its lawn sprinklers turned on. Family members returned to find their hillside lawn and home sitting in the valley below.

Other human actions may trigger mass movements as well. When we clear-cut forested slopes or accidentally set

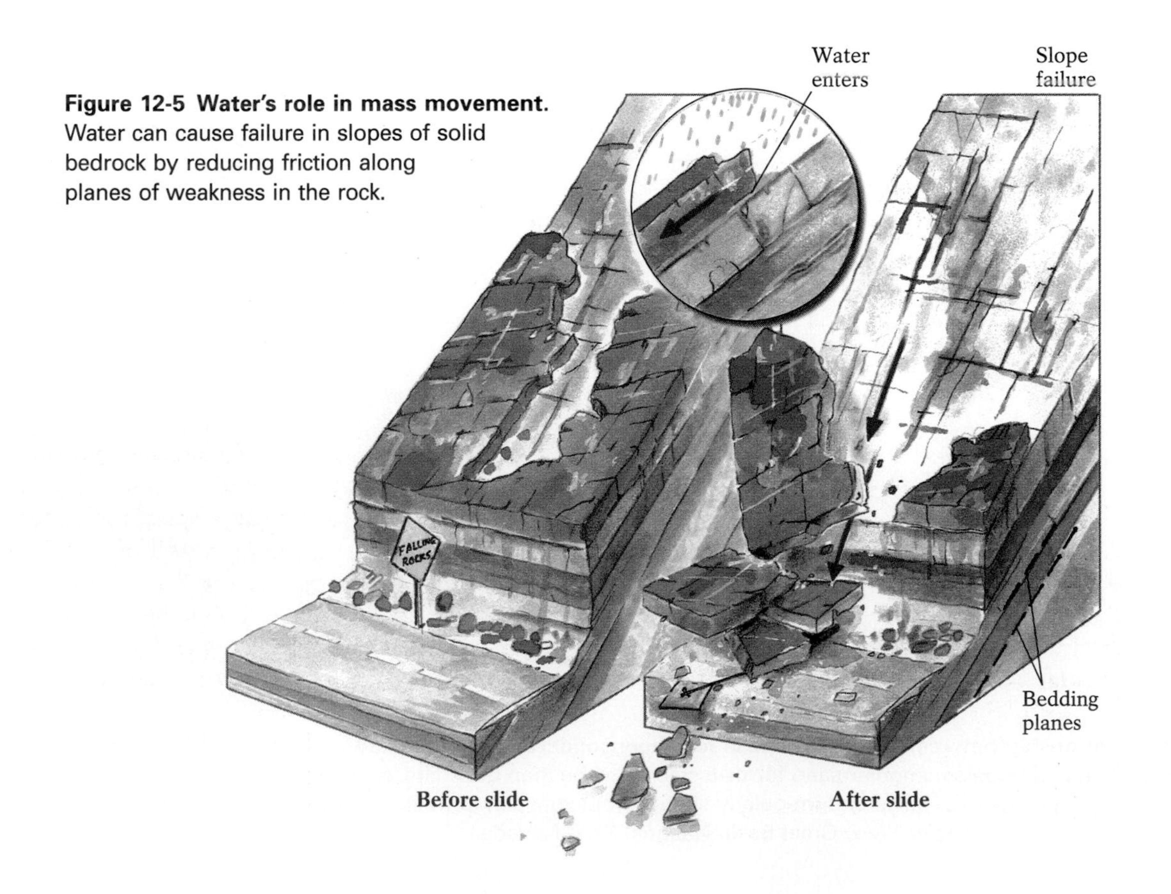

Figure 12-5 Water's role in mass movement. Water can cause failure in slopes of solid bedrock by reducing friction along planes of weakness in the rock.

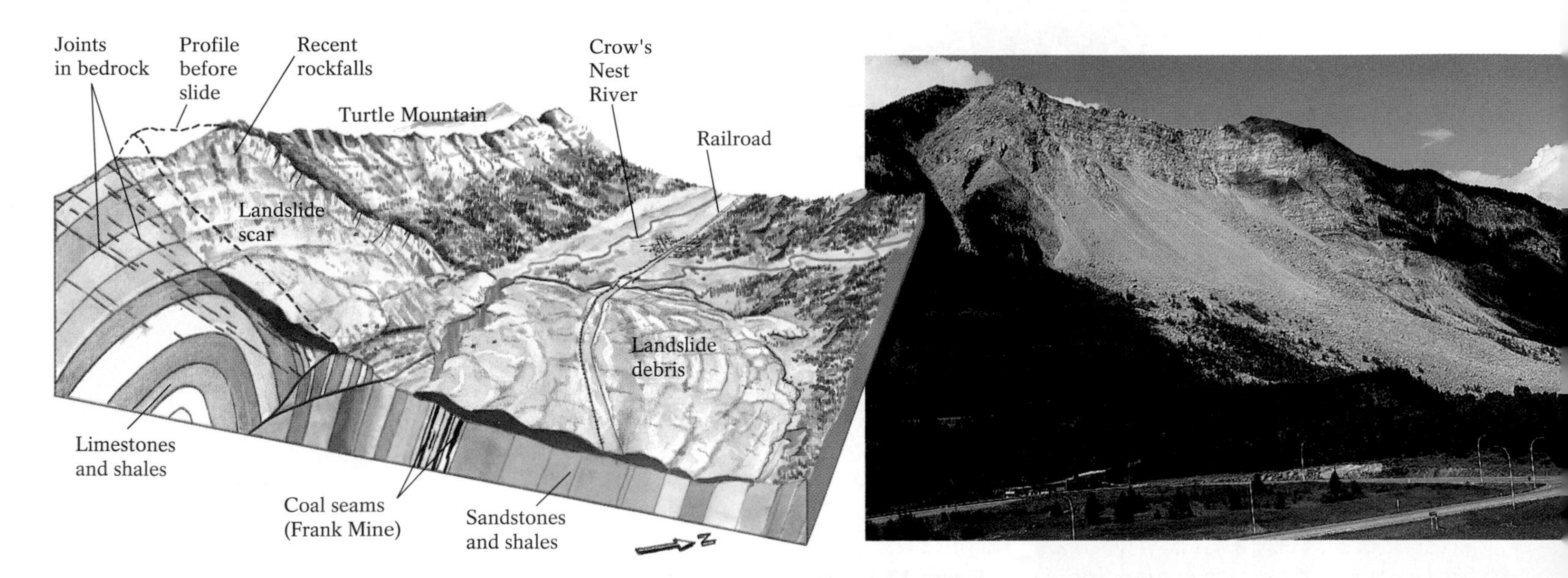

Figure 12-6 The 1903 Turtle Mountain landslide in Frank, Alberta. The disaster began when an enormous mass of limestone broke free and slid downslope to a protruding rock ledge, where it was launched airborne toward the valley below. After a 900-meter (3000-foot) drop, the rock struck the weak shales and coal seams in the valley, shattering into a great avalanche of crushed rock that spread at speeds as high as 100 kilometers (60 miles) per hour to a distance of 3 kilometers (2 miles) across the valley. The mass of crushed rock had such momentum that it actually ascended 120 meters (400 feet) up the opposite side of the valley. Ironically, 16 men working in the mines survived by digging their way out through a soft coal seam. Photo: Turtle Mountain as it looks today, showing evidence of more recent rockfalls within the scar from the 1903 slide.

forest fires, we eliminate the deep, extensive root networks that bind loose materials together. When we cut into the bases of sensitive slopes to clear land for homes or roads—especially if we dump the removed or other material on the top of the slopes—we oversteepen the slopes and jeopardize their equilibrium. Construction of housing developments on hillsides or cliffsides, as is common in California, can initiate mass movement when slopes bearing the extra weight of roads, buildings, and swimming pools later become weakened by heavy rainfall. Where extreme instability exists, even the vibrations of loud sounds—such as passing trains, aircraft sonic booms, and construction blasting—can trigger mass-movement events.

Sometimes human activities, such as mining, can combine with natural factors to increase the probability of mass movement along a slope. One such situation led to the Turtle Mountain landslide of 1903, near the Canadian Rockies town of Frank, Alberta, which claimed 70 lives (Fig. 12-6). The removal of a large volume of coal near the foot of the mountain weakened an already unstable structure containing numerous joints and fractures. The month preceding the slide had been a wet one in the southern Alberta Rockies, and water from the copious snowmelt probably entered and lubricated the fractures in Turtle Mountain, further increasing its potential for mass movement.

Types of Mass Movement

No universally accepted scheme exists for the classification of mass movements. Ask a soil scientist, a geologist, and a civil engineer to classify a mudflow (a rapidly flowing slurry of mud and water), and you'll likely get three different responses. Geologists, however, generally classify mass movements based on the speed and the manner in which the materials move downslope.

Slow Mass Movement

Creep, the slowest type of mass movement, is measured in millimeters or centimeters per year. It occurs virtually everywhere, even on the gentlest slopes, and generally affects unconsolidated materials, such as soil or regolith, whose depths rarely exceed a few meters. Loose material experiencing creep undergoes continuous rearrangement as individual particles become dislodged—for example, by burrowing animals and insects, raindrops, or swaying plants—and respond to the influence of gravity. Because of the friction between this surface material and the underlying slope material, a mass of material undergoing creep tends to move more rapidly at its surface and more slowly at deeper layers, causing implanted

13

Streams and Floods

On June 11, 1993, a foot of rain fell in southern Minnesota. Four days later, more than 11 additional inches drenched the same area. Thus began the wettest June, and North America's worst flood, since compilation of weather records began in 1878. Before the wet weather ended two months later, the upper Mississippi River had surged over its banks along an 800-kilometer (500-mile) stretch from St. Paul, Minnesota, to St. Louis, Missouri.

Life throughout a 12-state region was disrupted in countless ways. The swirling floodwaters undermined interstate bridges, washed away roads, and halted all barge traffic along the upper Mississippi, the economic artery of the Midwest. In Hannibal, Missouri, Mark Twain's boyhood home, children caught catfish on streets where they had ridden bicycles only days before. In Davenport, Iowa, ducks swam lazily in the 4 meters (14 feet) of water covering the outfield of Davenport Stadium. Two hundred fifty thousand Iowans in Des Moines went without drinking water for days after floodwaters contaminated the municipal water supply with raw sewage and chemical fertilizers. Rushing, muddy floodwaters—carrying uprooted trees and junked cars—swept away thousands of homes and businesses, displaced more than 50,000 people, took dozens of lives, and caused more than $10 billion in property damage and agricultural losses.

In addition to bringing floods (Fig. 13-1), rivers such as the Mississippi offer countless benefits. For example, they provide a steady supply of water for home and industrial use, transportation, recreation, and irrigation. Rivers can be used to generate clean electric power. Crops grown in the fertile soils near rivers provide nourishment for one-third of the Earth's human population. Historically, the success and prosperity of most major cities have been linked to the size of their rivers.

From the mighty Mississippi River to the smallest creek, any surface water whose flow is normally confined to a channel is called a **stream**, whether it is known locally as a river, creek, brook, or run. (Geologists use the terms *stream* and *river* interchangeably.) Streams are among the most

Figure 13-1 Flood damage from the April 1997 flood in Grand Forks, North Dakota.

Figure 14-1 Carlsbad Caverns in New Mexico. This cave was dissolved out of solid limestone bedrock by circulating groundwater.

14

Groundwater, Caves, and Karst

For thousands of years, wells and natural springs have supplied clean, abundant groundwater to human communities throughout the world. Even today, the presence of an adequate supply of uncontaminated groundwater can determine whether a region or community will grow and prosper. Pure water has even become a commercially valuable commodity, and both supermarkets and gourmet restaurants sell bottles of ordinary groundwater at exorbitant prices. This source of our trendiest beverage is also responsible for some of the world's most popular tourist attractions, dissolving spectacular caves out of solid bedrock (Fig. 14-1). Most caves form when acidic groundwater, flowing unseen beneath the Earth's surface, gradually enlarges tiny crevices in limestone to create huge, complex cave systems.

Groundwater accounts for 97% of the world's supply of unfrozen fresh water. It provides more than 50% of our drinking water, 40% of our irrigation water, and 25% of water used for industrial purposes. Throughout North America, we are withdrawing groundwater reserves that took thousands of years to accumulate, and supplies (particularly in the Southwest) are being depleted. Between 1955 and 1985, U.S. groundwater use doubled to keep pace with the population growth. In many places, however, groundwater has become contaminated because of improper disposal of wastes and other human activities.

Some of the most urgent issues facing the world's citizens relate to groundwater: where to find it, how to keep it clean, who owns it. In most areas, climate and local geology are capable of providing a continuing supply of cool, refreshing, healthy groundwater for present and future generations—if we learn to properly manage and preserve that reserve.

Groundwater Recharge and Flow

Groundwater *recharge* is the infiltration of water, mostly from precipitation, into the Earth's groundwater systems. After infiltration, groundwater flows through soils and rocks until it

15

Figure 15-1 The deeply crevassed Davidson Glacier near Haines, Alaska.

Glaciers and Ice Ages

Glaciers provide some of the most awe-inspiring scenery on Earth, drawing millions to Alaska and other high-latitude regions in which they are common (Fig. 15-1). Glaciers also shape the landscapes over which they slowly flow. Without the continental ice sheets of the Pleistocene Epoch, which ended approximately 10,000 years ago, North America would not have its Great Lakes, Niagara Falls, Hudson Bay, Puget Sound, or the 15,000 lakes of Minnesota. There would be no Cape Cod in Massachusetts and no fertile rolling hills in the Midwest and southern Canada. Rivers such as the Missouri and Ohio would drain north to the Arctic and Atlantic Oceans, rather than south to the Mississippi River and the Gulf of Mexico. If Earth had no glaciers today, the shapes of the continents themselves would be substantially different, because sea level would be nearly 70 meters (230 feet) higher than its current level. Landlocked cities such as Memphis, Tennessee, and Sacramento, California, would be seaports, while San Francisco, New York, and many other coastal cities would be mostly under water.

Today, glaciers cover about 10% of the world's land surface. At very high latitudes, such as in the Arctic, Antarctica, and Greenland, these features are common and can exist at any elevation, even at sea level. They occur in mid-latitude regions as well, although only where temperatures are sufficiently cold—at high elevations (at about 2500 to 3000 meters [8000 to 10,000 feet]), such as in the Northern Rockies of Alberta, British Columbia, and Montana.

In this chapter, we discuss what glaciers are, where they exist today, where they appeared in the past, and how they have shaped the landscape. We also consider why only certain periods of the Earth's history have been marked by worldwide glacial expansions, and speculate about when glaciers may again cover vast areas of North America.

Glacier Formation and Growth

A **glacier** is a moving body of ice that forms from the accumulation and compaction of snow. Glaciers flow downslope or outward under the influence of gravity and the pressure of their own weight.

Glacier formation begins with a snowfall and the accumulation of snowflakes (Fig. 15-2). The initial snowpack, like most new snowfalls, tends to be fluffy and of low density. As more snow falls, however, it compresses the underlying snow. Some of the contact points between snow crystals then melt. The resulting water migrates into the pore spaces between snow crystals, refreezes, and binds the snow crystals together. The dense, well-packed snow, called **firn,** created in this way can survive the summer melting season. Repeated melting and refreezing of its interlocking crystals eventually squeezes nearly all air from firn, allowing it to become glacial ice, with a density about the same as that of an ordinary ice cube.

Glaciers are able to form and grow when snow accumulation exceeds the losses from summer melting and sublimation. (*Sublimation* is the process in which a solid changes directly into a gas; in this case, some of the solid snow changes directly into water vapor. This sublimation process accounts for decreases in snow that occur without a "slush" stage.) Cool, cloudy, brief summers that minimize melting probably contribute more to glacier growth than very low winter temperatures; in places where summer warmth completely removes even heavy winter accumulations of snow, such as in Minnesota, glaciers do not occur. The climates that are most likely to produce glaciers are found in higher latitudes and at higher elevations, where it rarely remains very warm for very long. In mountainous regions, glaciers usually form at or above the *snowline,* the lowest topographic level at which we find year-round snow.

The time required for fresh snow to become transformed into glacial ice varies with the rate of snow accumulation and the local climate. In snowy climates with an average annual air temperature close to the melting point of ice (0°C [32°F]), snow may be converted to ice in only a few decades. In extremely cold polar settings, where snowfall is typically sparse and little melting takes place, glacial ice formation may take thousands of years.

Classifying Glaciers

Glaciers are classified broadly by whether the local topography confines them or allows them to move freely (Fig. 15-3). **Alpine glaciers** are confined by surrounding bedrock highlands. Because they are confined, they are relatively small. Three types of alpine glaciers exist: **Cirque glaciers** create and occupy semicircular basins on mountainsides, usually near the heads of valleys; **valley glaciers** develop when growing cirque glaciers flow into preexisting stream valleys; and **icecaps** form at the tops of mountains.

Piedmont ("foot of the mountain") glaciers originate as confined alpine glaciers but then flow onto adjacent lowlands; once unconfined in the lowlands, they can spread radially. Piedmont glaciers that flow to coastlines and into seawater become *tidewater glaciers.*

The only completely unconfined form of glacier is a **continental ice sheet,** an ice mass so large that it blankets much or all of a continent. Today, continental ice sheets cover Greenland and Antarctica. The Antarctic ice sheet actually consists of two ice sheets separated by the Transantarctic Mountains. It is as much as 4.3 kilometers (nearly 3 miles) thick in places and occupies an area about 1.5 times as large as the continental United States. Twenty thousand years ago, such vast ice sheets covered North America to south of the Great Lakes and blanketed western Europe south to Germany and Poland.

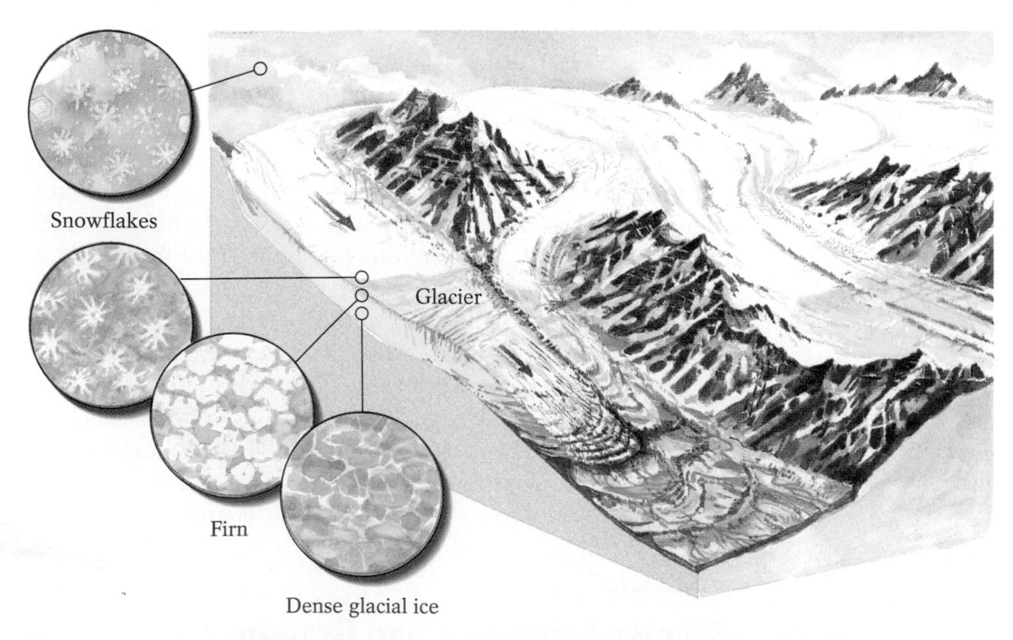

Figure 15-2 The formation of glacial ice from snow. Hexagonal snowflakes that fall on a glacier are gradually changed into rounded crystals. As overlying snow buries the crystals and exerts pressure on them, they become packed into increasingly dense firn and finally glacial ice.

Figure 15-3 Types of glaciers. **(a)** Angel Glacier, a cirque glacier on Mount Edith Cavell, Jasper National Park, Canada. **(b)** A valley glacier in Tongas National Forest, Alaska. **(c)** An icecap in the Sentinel Range, part of the Antarctic continental glacier. **(d)** A tidewater glacier at Kenai Fjords National Park, Alaska.

16

Deserts and Wind Action

Unlike any landscape described earlier in this book, Death Valley National Monument, located in southeastern California, displays the wind-swept sands and austere barrenness characteristic of an arid landscape (Fig. 16-1). The valley owes its unique appearance to its extreme lack of water. Without appreciable water, little chemical weathering can take place—soils are thin, dry, and crumbly, and winds readily sweep loose particles into dunes or sandblast exposed rock surfaces with them. These characteristics are typical of **deserts,** regions that receive very little annual rainfall and are generally sparsely vegetated.

Every major continent, although surrounded by water, contains at least one extensive dry region. Deserts account for as much as one-third of the Earth's land surface—more area than is occupied by any other terrestrial environment. Relatively few deserts, however, resemble the popular image of endless tracts of drifting sand. In North Africa's Sahara, the world's largest desert, only 10% of the surface is sand-covered. Even the Arabian Desert, Earth's sandiest, is only 30% sand-covered. Desert climates are *arid.* That is, they are characterized by dryness rather than by temperature, and thus they can be found in both cold and hot regions. Such diverse landscapes as ice-bound Antarctica, the fog-shrouded coasts of Peru and Chile, and the near-continuous 8000-kilometer (5000-mile) stretch of land across northern Africa and the Arabian peninsula to southern Iran are all considered deserts, based on their extreme lack of surface water.

The term "desert" is misleading in its implication that the land is literally deserted, devoid of life. In fact, hot deserts are home to some of the Earth's hardiest plants and animals (Fig. 16-2), which have developed special adaptations that enable them to survive in extremely dry conditions. Some hot-desert plants produce seeds that can endure 50 years of drought. Some have small, waxy leaves that minimize water loss to evaporation. Most possess thick, spongy stems that store water from the occasional cloudburst and produce deep root systems through which to tap groundwater supplies. Some desert plants resemble dead twigs for months or even years on end, until the occasional downpour arouses them into a brief but memorable bloom.

Figure 16-1 Wind-swept sand dunes at Mesquite Flats, Death Valley National Monument, California.

17

Shores and Coastal Processes

North Americans have a passion for vacationing near shores and coasts, and more than half of us live within 80 kilometers (50 miles) of the Atlantic or Pacific Ocean or near one of the Great Lakes. Shores and coasts, at once scenic and educational, are wonderful places to observe natural processes—particularly the action of waves, tides, and nearshore currents.

All shorelines change constantly through natural processes. Sometimes those processes act rapidly and dramatically. Waves from a powerful storm, such as those shown in Figure 17-1, may produce immediate alterations to the shoreline. On January 2, 1987, for example, waves from a powerful winter storm gouged more than 20 meters (65 feet) of dunes and beaches from Nauset Beach, at the bend in the "elbow" of Cape Cod along the Atlantic coast of Massachusetts. The beach, a 20-kilometer (12-mile)-long pile of sand that had been built and shaped by waves and tides during the past 4000 years, had sheltered the bayside town of Chatham and its fishing fleets for centuries. On that night, however, Nauset was breached by 6-meter (20-foot)-high waves that also swept away nearly a kilometer of the cape's offshore islands, the barrier that had protected Chatham from the Atlantic's waves and storms.

The entire land region bordering a body of water is called a **coast**. Coasts extend inland until they encounter a different geographical setting, such as a mountain range or a high plateau. A **shoreline** is the boundary where a body of water meets the adjacent dry land. A *shore*, the strip of coast closest to a sea or lake, often includes a sandy strip of land, or a *beach*. In this chapter, we examine the variety of processes that shape and change our coasts—the waves, currents, and tides that erode and deposit coastal materials. We also look at several different types of coasts and consider how human activity sometimes affects the evolution of coasts.

Waves, Currents, and Tides

Moving water is the great agent of geologic change at the Earth's coasts. Water can be set in motion by the wind, which

Figure 17-1 Crashing surf along the central Oregon coast.

18

Human Use of the Earth's Resources

Until recently, the Earth's resources, such as those being mined in Figure 18-1, were believed to be unlimited. Yet, today, we face serious shortages of many essential materials. For example, scientists believe that the world's recoverable supply of crude oil, from which we obtain gasoline, may last only another 50 to 100 years at the current rate of use. How have we managed to exhaust our stores so quickly? Can we compensate for these losses?

To find the answers to these questions, we must understand the growth rate of the world's population, the amount of natural resources used by each individual, and the search for alternative resources. In the United States alone, each person directly or indirectly uses about 10,000 kilograms (22,000 pounds) of raw materials each year; most of this amount consists of stone and cement for the construction of roads and buildings, but it also includes about 500 kilograms (1100 pounds) of steel, 25 kilograms (55 pounds) of aluminum, and 200 kilograms (440 pounds) of industrial salt (mostly for cold-weather road maintenance). Each American also uses nearly 3800 liters (1000 gallons) of oil per year. Collectively, Americans account for approximately 30% of world oil consumption. The United States, which possesses only 6% of the world's population, uses nearly 30% of its minerals, metals, and energy. One American uses as many as 30 times as much material and energy as a person in an emerging nation.

Resource consumption worldwide is rising at an accelerating rate as world population increases (it is now approaching 6 billion—three times what it was in 1920—and expected to double by 2040) and people everywhere strive to achieve the benefits of technological development. Unless we identify new supplies of depleted resources or find substitutes for them, and until we manage industrial development in ways that limit resource depletion, impending shortages will force people everywhere to change their ways of life.

Figure 18-1 The Bingham Copper Mine. The largest open cut copper mine in North America, it has yielded more than $5 billion worth of copper.

Reserves and Resources

Reserves are natural resources that have been discovered and can be exploited profitably with existing technology and

19

Interpreting the Past

"Do nothing to mar its grandeur," declared President Theodore Roosevelt, in conferring the status of national monument on the Grand Canyon, which he described as "the one great sight which every American should see." What is it about the vistas that greet even the most casual visitor to the canyon rim that inspire such awe? Could it be the sense that from the precipice, one is actually peering back into the abyss of time? The Grand Canyon was still uncharted wilderness when John Wesley Powell led the first expedition down the Colorado River in 1869. Powell, a Union army major who lost most of his right arm in battle, was a school teacher with a keen interest in natural history. He also was undoubtedly the only army commander who collected fossils as his troops dug trenches in preparation for battle. Running the river rapids in wooden boats, Powell and his men endured considerable peril and great physical hardship. Yet in the midst of adversity, Powell found himself fascinated by the canyon walls that surrounded him, noting in his diary, "All about me are interesting geologic records, the book is open and I read as I run." Powell, who would later become the director of the United States Geological Survey, recognized that something special was written on the walls of the canyon—the story of Earth's history (Fig. 19-1). We can now read the same story as Powell by applying the principles learned in the first three parts of this book to the rocks we see around us everywhere, not just in the Grand Canyon. The final eight chapters of this text describe how we have interpreted the history of geologic change and of the evolution of life on Earth.

This first chapter describes the methods we use to interpret the geologic past. In the nineteenth century, geologists began to understand the history of the Earth, and of life on Earth, as a sequence of events that could be interpreted from the study of the geologic record. They were equipped for this task with the principle of uniformitarianism, first developed by James Hutton in the late eighteenth century (see Chapter 1). The application of this principle was vigorously promoted by Charles Lyell, a British geologist who published the textbook *Principles of Geology* in 1830. Applying the principles of relative dating to make correlations (see

Figure 19-1 The Grand Canyon. The view from the rim of the Grand Canyon permits visitors to observe hundreds of millions of years of geologic history. Rock layers ranging in age from Precambrian to late Paleozoic were exposed by downcutting of the Colorado River during the last 2 million years.

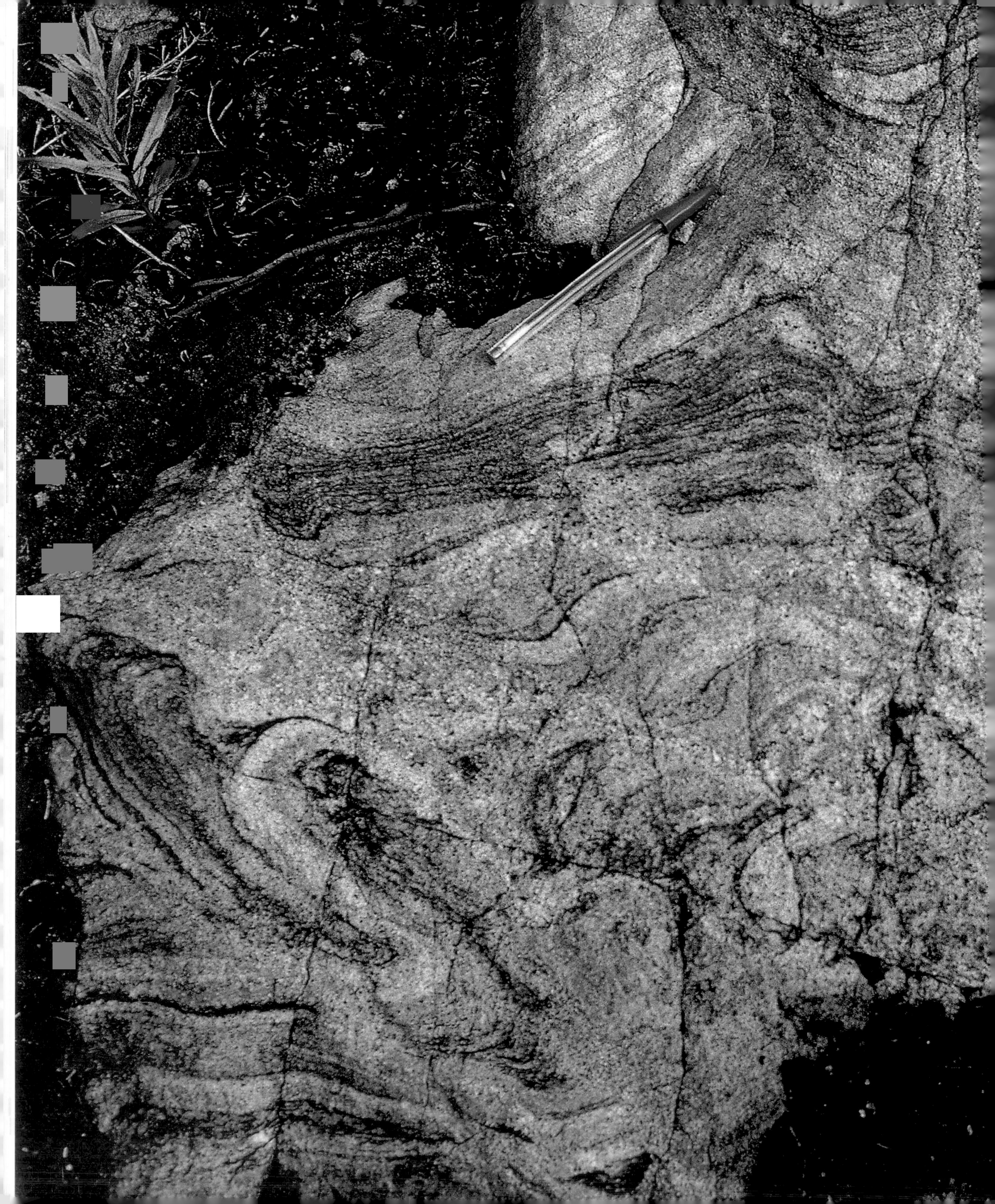

20

The Precambrian Record

What are the oldest rocks found on Earth? The answer to this particular question was long in coming. Estimates of the age of the Earth (and therefore the maximum age of its rocks) by nineteenth century scientists were little more than educated guesses, providing crude estimates ranging from a few million to over a billion years. A method of directly measuring the ages of rocks awaited the discovery of radioactivity in 1896 by the chemist Henri Becquerel. The true antiquity of Precambrian rock formations, which provide us with glimpses into Earth's earliest history, was finally demonstrated following the development of radiometric dating techniques in the early twentieth century. Measurements of the ages of Precambrian rocks quickly confirmed the most extreme estimates of the age of the Earth, establishing an existence measured in billions of years. Strictly speaking, the answer to the question that began this page would be the nearly 4.6-billion-year-old meteorites that have landed on our planet. As explained in Chapter 8, these extraterrestrial rocks are remnants from the formation of our solar system that allow us to fix a date for the origin of all of the bodies in the solar system. Given an age of 4.6 billion years for our planet, can we find any rocks of this age on Earth?

Because tectonic processes (subduction in particular) destroy the crust of the oceans, the very oldest rocks must be located in the continents. During the late twentieth century, geologists devoted considerable energy to examining the shield areas of the continents (the exposed portions of the cratons) in pursuit of the most ancient pieces of Earth's crust. By the 1980s, rocks from various areas in the Canadian Shield, as well as regions of South Africa and Australia, were identified as having ages of about 3.5 billion years. The trail eventually led to western Greenland, where rocks are found that date to 3.8 billion years, and similar ages have since been found for formations in several locations in North America, Antarctica, Australia, and China. The distinction of the oldest known crust belongs, however, to one particular formation found in the Northwest Territories of Canada. The Acasta Gneiss, a high-grade metamorphic rock with a nearly felsic composition (Fig. 20-1), contains crystals of the mineral

Figure 20-1 Acasta Gneiss. This metamorphic formation found in the Northwest Territories of Canada is the oldest known formation of crustal rocks. Zircons in the formation have been dated to 3.96 billion years.

21

The Geology of the Paleozoic Era

Picture warm, clear ocean waters lapping peacefully onto white, sandy beaches while coral reefs, teeming with life, lie submerged offshore. You might (with considerable justification) associate these images with some modern tropical paradise at low latitudes. During the early Paleozoic, this scene described locations now far removed from the tropics; places such as northern Canada, Greenland, and Siberia lay in the tropical latitudes of an Earth that looked much different than it does today. When the Paleozoic Era began, the fragments of the Proterozoic supercontinent Rodinia were still drifting apart but were mostly located near the equator. The geography was not the only important difference between the ancient and modern worlds; the world of the very early Paleozoic Era would appear strangely empty and quiet to us, lacking palm trees, tropical birds, or biting insects, for the landscape was entirely barren of plant and animal life. But this scene changed dramatically as the Paleozoic witnessed the appearance of most major plant and animal groups on the lands and in the oceans (the subject of the next chapter).

The quartz-rich sands that were deposited in the shallow seas at the dawn of the Paleozoic resulted from a sea-level rise that eventually covered most of the North American craton (Fig. 21-1). This transgression was just the first of a series of major transgressions and regressions that are recorded very clearly in the stratigraphic record. Additionally, the tectonic quiet that marked the start of the Paleozoic Era was not long lived. Continuation of the Wilson cycle of tectonic events, begun by the breakup of Rodina, led to a series of collisions that culminated in the assembly of the new supercontinent Pangaea (named by Alfred Wegener). As we shall see, these continental collisions were also responsible for building some impressive mountain belts, including the Appalachians.

Figure 21-1 Shallow marine sandstone. The quartz-rich Potsdam Sandstone, seen here near Ausable in northern New York, was deposited near the shore of a shallow sea that covered most of the North American craton early in the Cambrian Period.

The Divisions of the Paleozoic

The Phanerozoic Eon (the time of "visible life") opens with the Paleozoic Era, the time of "ancient life." In North

22

Paleozoic Life

Figure 22-1 Burgess Shale. Charles Walcott made his discovery in the Burgess Shale in this mountain pass near Mount Wapta, now in Yoho National Park, British Columbia.

The discovery was made one day at the end of the summer of 1909 as Charles Walcott and his party, including his family, were riding along a ridge near Mount Wapta, in the Canadian Rockies of British Columbia (Fig. 22-1). Walcott at this time was secretary of the Smithsonian Institution and the foremost expert on the geology of the Cambrian Period, which was well exposed in the rocks of this mountain pass. According to an account written after Walcott's death, the party was descending the trail in a snowstorm when the horse Walcott's wife was riding slipped on rocks in the trail. Walcott stopped to clear a slab of rock from the path and found fossils unlike any he had seen previously. Although this account is probably somewhat embellished (Walcott's own journal shows that the first of these fossils was found several days earlier), the fact remains that Walcott and his party did indeed find slabs containing exquisitely preserved fossils of soft-bodied organisms, many quite bizarre, of Middle Cambrian age. Walcott immediately recognized the importance of this find, but the slabs had fallen from higher slopes that were already snow covered. Walcott returned the following year and successfully traced the rock slabs back to a layer in the Burgess Shale from which they had fallen, establishing a quarry that yielded about 60,000 specimens over the course of several seasons. Walcott stored and studied the collection at the U.S. National Museum of Natural History of the Smithsonian Institution. He catalogued this collection of fossils, most of which were previously unknown, and attempted to classify the organisms by placing them within known animal categories. But many years later, paleontologist Harry Whittington undertook an exhaustive examination of these fossils and concluded that many actually represent completely unknown groups of organisms, new phyla that never had been described before.

What makes this collection of fossil specimens so particularly exciting is the stunning diversity of forms, including an amazing 170 species. The Burgess Shale fauna, as it is now known, has given us a window on the Cambrian world, showing us that life was quite complex by Middle Cambrian time. Why should we find this complexity of Cambrian life so

Graptolites Another important fossil group for stratigraphic correlation of portions of the Paleozoic record are the **graptolites,** a class of the *hemichordate* phylum. Hemichordates, a small group including the modern acorn worms, are organisms with a threefold body division and central nerve cord. The carbonized impressions of graptolites, which appear like small saw blades, are the remains of colonies of small filter-feeding animals that lived in cups 1 millimeter (0.04 inch) long, arranged in rows along one or both sides of an elongated stem (Fig. 22-17). Clusters of these stems formed upward-branching structures anchored to the sea floor during the Cambrian but graptolites evolved free-floating forms by the Ordovician. The planktonic colonies were spread easily by ocean currents and are commonly preserved in the deposits of deep oceanic environments worldwide. Graptolites evolved rapidly, which, combined with their widespread distribution, makes them particularly useful for stratigraphic correlation of rocks of Ordovician and Silurian age. They declined following the Early Devonian Period, however, and became extinct during the Carboniferous.

(a)

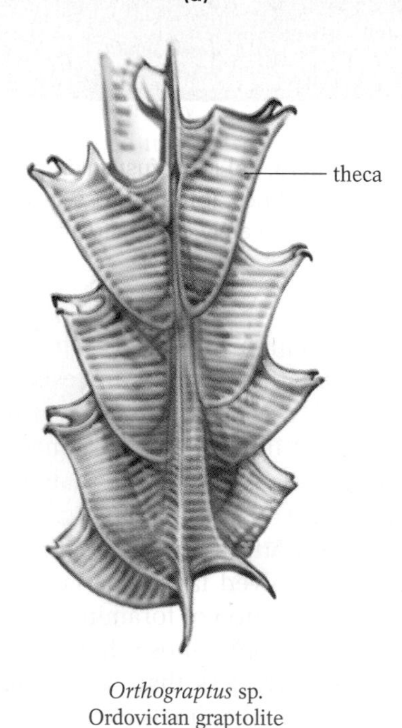

(b)

Figure 22-17 Graptolites. **(a)** The sawbladelike impressions are the carbonized remains of the Silurian graptolite *Monograptus*. **(b)** The individual animals in a graptolite colony lived in the tiny, cuplike thecae.

Paleozoic Reefs What visual images are conjured by the word *reef?* You probably picture great walls of brightly colored corals in a vast array of shapes, perhaps surrounded by schools of brilliant tropical fish. Actually, modern reefs are complex ecological systems, hosting a myriad of species, but corals are indeed an important part of most reef systems. They form a rigid framework that, by being wave resistant, provides a sheltered environment for many other organisms. Paleozoic reefs were also complex ecological systems, but they changed with time as different organisms evolved to build this rigid framework. And just as today, many other benthic organisms, such as crinoids and brachiopods, took advantage of the sheltered environment provided by reef structures, contributing to the accumulation of sediment in these settings.

Corals were actually not the original reef builders of the Paleozoic Era, being preceded in this role by other organisms. The first organic reef structures were built by clusters of cone-shaped animals called **archaeocyathids** ("ancient cups") (Fig. 22-18). Most paleontologists now classify the archaeocyathids as sponges (Phylum Porifera), although others place them in a separate phylum. These calcareous cups grew anchored to the sea floor in shallow environments and covered extensive areas with their skeletons. They appeared during the Early Cambrian, coinciding with the dominance of the small shelly fauna, and quickly diversified into about 200 species. Archaeocyathids almost disappeared at the end of

Figure 22-18 Archaeocyathids. Clusters of archaeocyathids, shown in growth position, formed the earliest reefs.

the Early Cambrian Period, however, and were completely extinct by the end of the Late Cambrian.

Corals are members of the *cnidarians,* a phylum that consists of soft-bodied animals possessing tentacles equipped with stinging cells; other common modern cnidarians include jellyfish, sea anemones, and the hydra, none of which produce a rigid enclosing structure as do corals. The corals we are familiar with from modern reefs (called *scleractinians*) did not evolve until the Mesozoic Era, but other orders of corals were important in Paleozoic reefs. The **tabulate corals,** which appeared during the Cambrian, were colonial organisms that built fairly simple chain and honeycomb-like structures (Fig. 22-19a). These corals, such as the common *Halysites,* were prominent in reefs from the mid-Ordovician through Silurian periods, following which they gradually declined. A change in the composition of Paleozoic reefs resulted from the rising prominence of **rugose corals.** Often called horn corals, the larger rugose coral polyp built vase-shaped structures that stood individually or combined to form colonies (Fig. 22-19b). Rugose corals were particularly common during the Devonian and Carboniferous periods but became extinct at the end of the Permian Period, as did the tabulate corals.

Another group of calcareous sponges, the **stromatoporoids,** built layered and mound-shaped structures, some as wide as 5 meters (16 feet), from laminar sheets of calcite. These organisms contributed greatly to the growth of reefs through much of the Paleozoic Era (Fig. 22-20); by growing over and binding together the broken remains of other reef dwellers, they served to solidify the framework of the reef,

(a)

(b)

Figure 22-19 Paleozoic corals. (a) *Halysites* was a common tabulate coral of the Ordovician and Silurian. **(b)** *Heliophyllum* was a solitary rugose coral that is well known from the Middle Devonian.

Figure 22-20 Stromatoporoids. Many Devonian reefs were built largely by stromatoporoids with cabbagelike (as seen here) or sheetlike shapes.

Figure 22-30 Earliest seedless vascular plant. The fossil *Cooksonia,* of Late Silurian age, is the earliest known seedless vascular plant.

development of leaves also increased the efficiency of photosynthesis by increasing the surface area for catching sunlight. The formation of root systems provided stronger support for growth and increased the capacity for collecting water and nutrients.

Seedless Vascular Plants

These early plants were **seedless vascular plants**—that is, they reproduced by means of spores, as modern ferns do; the spores germinate to grow into a special form of the plant, the gametophyte, which produces the male and female gametes. Because the sperm of the gametophyte requires moisture to fertilize the eggs, these plants were restricted to moist habitats. Even with this environmental restriction, seedless vascular plants quickly diversified into several groups. During the Late Devonian Period, some of these spore-bearing plants formed extensive forests of trees reaching heights of over 30 meters (100 feet). The growth of these forests may have been crucial in controlling the composition of the Paleozoic atmosphere. During the Carboniferous Period, use of carbon dioxide for photosynthesis and burial of the plants in coal swamps caused atmospheric concentrations of this gas to plunge, contributing to the onset of the late Paleozoic glaciation of Gondwana. Simultaneously, oxygen levels soared to perhaps as high as 30%, compared to the present value of 21%.

In addition to ferns, three important groups of seedless plants emerged during the Devonian Period. The **lycopsids,** which are represented today by the rather inconspicuous club mosses, appeared by the Early Devonian or even earlier. During the Carboniferous, the lycopsids included trees as tall as 30 meters (100 feet) that dominated swampy habitats. These trees, including *Lepidodendron,* are sometimes called scale trees for the pattern of scars on the trunk formed by the attachment of the leaves (Fig. 22-31a). Another important group of seedless vascular plants, the **sphenopsids,** appeared by the Late Devonian. These plants are characterized by their jointed stems and are represented today by a single genus that contains horsetails and scouring rushes. The most common Paleozoic sphenopsid was *Calamites* (Fig. 22-31b); it grew abundantly along river banks, commonly to heights of 5 or 6 meters (16–20 feet) but sometimes as high as 30 meters (100 feet). Along with the lycopsids and ferns, sphenopsids were abundant in the Pennsylvanian coal swamps; because lycopsids and sphenopsids both reproduce by spores, their distribution was largely limited to moist environments. A third group, the *progymnosperms,* so named because they may have been ancestral to the later gymnosperms, emerged during the Middle Devonian. *Archaeopteris,* the best-known example of this group, was a tree with a woody structure like modern trees that grew to heights of 35 meters (115 feet), but sported fernlike leaves and was spore bearing (Fig. 22-32).

Seed Plants

The development of seeds was a major advance in plant evolution. By producing a fertilized seed, rather than spores, plants were finally permitted to grow in any habitat in which the seed could germinate; in essence, plants were free of the

(a)

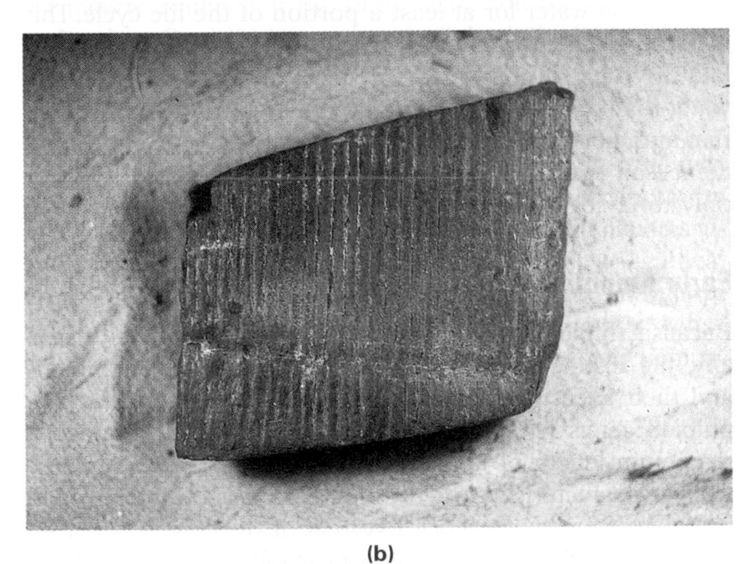

(b)

Figure 22-31 Paleozoic seedless plants. Common plants in the Pennsylvanian coal swamps included **(a)** the lycopsid *Lepidodendron,* and **(b)** the sphenopsid *Calamites.*

Figure 22-32 *Archaeopteris.* The impressions of this progymnosperm are common in the Devonian-age Catskill Formation of Pennsylvania and New York.

requirement for excessively moist environments and could spread across dry land. This evolutionary feat is first recognized in a group of plants that appeared during the Late Devonian Period called *seed ferns,* named for the fernlike pattern of the leaves. One late Paleozoic group of seed ferns, *Glossopteris,* was tree sized and widespread across Gondwana; the fossils of this plant were a clue to Alfred Wegener about the position of the continents at the end of the Paleozoic Era (Fig. 22-33). Also spreading across the late Paleozoic landscape were the first **gymnosperms,** the plant group that includes conifers. One group of early conifers (the *cordaites*) formed forests of trees that grew as tall as 50 meters (164 feet). The dominance of gymnosperms in the flora of the Permian Period can be attributed to the ability of these seeded plants to grow in much drier climates at a time when aridity became widespread.

The Arthropod Invasion

The first animals to make the transition to a terrestrial environment were arthropods. The remains of centipedes and arachnids (the class including spiders, scorpions, and mites) are found in rocks of Late Silurian age, and fossils of flightless insects (springtails) occur in the Lower Devonian strata of Scotland; these fossils demonstrate that the invasion of the land began soon after colonization by plants. Insects in particular underwent spectacular adaptive radiation during the late Paleozoic, with many winged forms, including mayflies

23

The Geology of the Mesozoic Era

About 75 million years ago, the vast seaway that covered most of western North America gradually shallowed. Sandy shorelines and barrier islands replaced the deep muddy waters, leaving behind massive sandstone layers that overlie a thick sequence of shale. Today, the sandstones, being more resistant to erosion, form the cliffs that support the top of Mesa Verde, a series of connected, flat-topped mountain ridges in southwestern Colorado bordered by the Mancos and Montezuma valleys. In many places, water seeping through the porous sandstone ledges dissolved the calcite cement that holds the grains together, weakening the rock and causing caves and alcoves to form in the cliff faces. The mesa and caves were inhabited by a group of ancient Americans (known as the Ancestral Puebloans) who moved into the region about 1500 years ago. Because the cooler climate at this high elevation allowed the soil to remain more moist than in the surrounding valleys, these ancient people found the high mesa top a more suitable location for farming and hunting. Eventually, the inhabitants of the mesa built elaborate stone and masonry dwellings, even entire villages, within the sheltered and easily defended cliff sites (Fig. 23-1).

The settlements were abandoned for unknown reasons about 700 years ago (long before European colonization); perhaps prolonged drought or overuse of natural resources reduced the ability of the land to support the population. For whatever reason, the inhabitants moved south to other areas of the Southwest, and the cliff dwellings were forgotten until they were rediscovered in the 1880s by cowboys searching for lost cattle. Although archaeologists have spent decades piecing together details of the lives of the Ancestral Puebloans, whose dwellings are now visited by tens of thousands yearly in Mesa Verde National Park, many questions remain about these ancient builders.

Mesa Verde National Park is just one small piece of the picturesque landscape of the great American West. With its lofty mountains, multihued vistas, and sculpted canyons, the West evokes the old axiom, "nothing is constant but change." Much of this scenery, exquisitely exposed in many of the national parks and monuments of the western states, was born

Figure 23-1 Cliff dwellings in Mesa Verde National Monument. Cretaceous sandstones of the Mesa Verde Group overlie the Mancos Shale in southwestern Colorado. Caves in the sandstone cliffs were inhabited by ancient Americans until about 700 years ago.

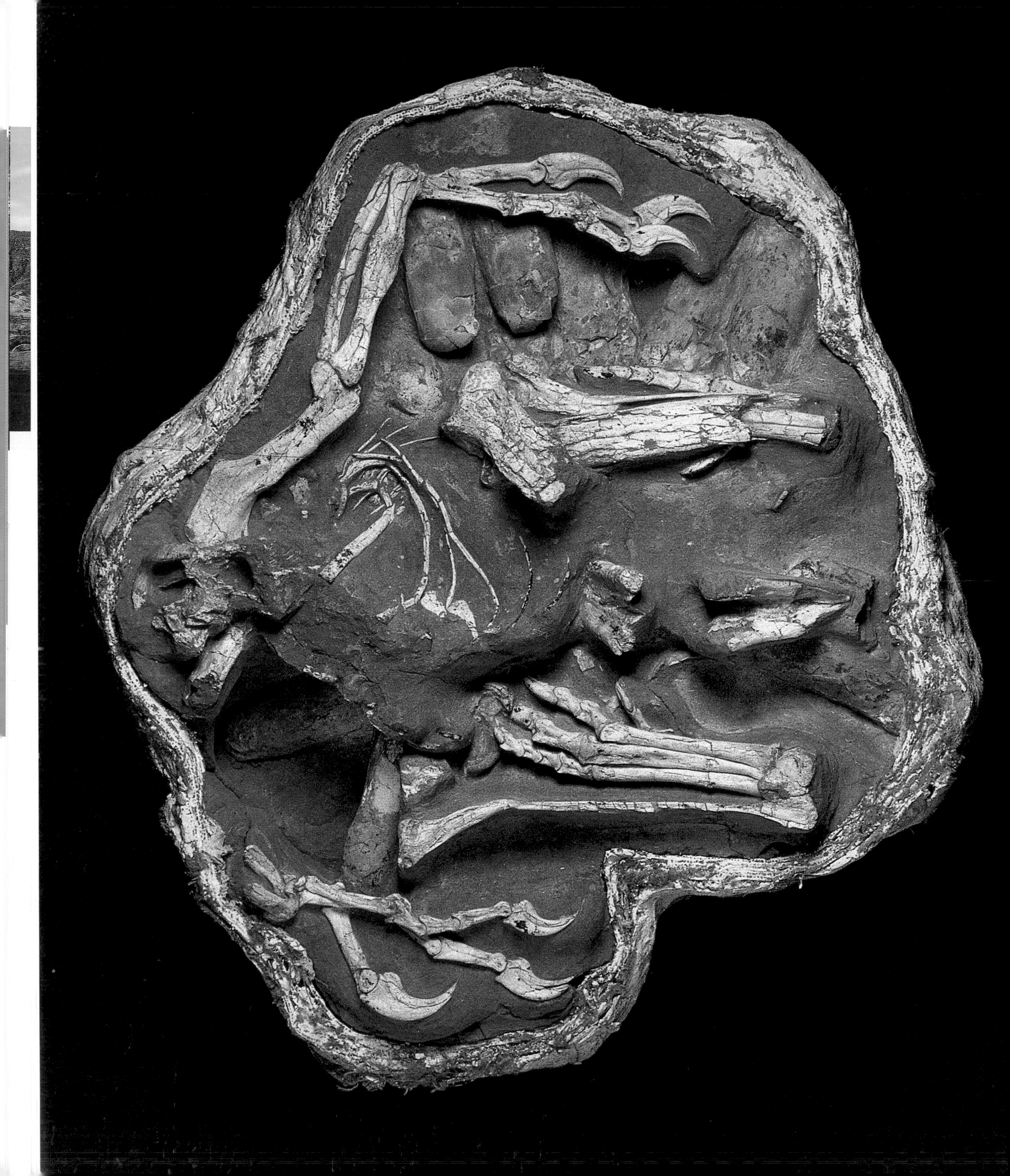

24

Mesozoic Life

Figure 24-1 Dinosaur eggs and the egg thief. This *Oviraptor* skeleton was discovered surrounding a nest, apparently in position to brood the eggs.

In 1922, Roy Chapman Andrews, a scientist on the staff of the American Museum of Natural History in New York, led the first of a series of expeditions to the Gobi Desert of Mongolia. The objective of the initial excursion was a search for the fossil remains of human ancestors (the so-called missing link), thought by some scientists of the day to have evolutionary roots in central Asia. Although Andrews's expedition covered vast areas of harsh desert terrain with cars and camels, it failed to recover fossils of primitive man. But other discoveries erased this disappointment. The remains of Cretaceous dinosaurs found by Andrews's team justified planning a return trip the following year, in particular to a location they named the Flaming Cliffs for the spectacular sunsets they witnessed there. On their first visit to this site, the team had found fragments of fossilized eggshells, which they assumed belonged to large birds. Greater discoveries awaited their arrival at this location the following year.

On returning in 1923, the team uncovered a treasure trove of dinosaur skeletons, including over 100 individuals, both adults and juveniles, of the early horned dinosaur *Protoceratops*. More amazing, the team also found nests of eggs that they now recognized as belonging to dinosaurs rather than birds, the first dinosaur eggs ever found. The eggs had been laid in carefully arranged circular patterns but had never hatched. Given the abundance of *Protoceratops* skeletons in the vicinity, the eggs were naturally presumed to belong to this herbivorous dinosaur, even though the bones of a small carnivorous dinosaur were actually found on one nest. This dinosaur was assumed to have died in the act of stealing the eggs, and so was given the name *Oviraptor* ("egg thief"). *Oviraptor* was exonerated in 1993 when another nest of identical eggs was discovered, again in the Gobi Desert, and again with an adult *Oviraptor* on the nest. This time, the nest contained the tiny fossil of an embryonic dinosaur identified as a baby *Oviraptor*. It would seem that in the initial rush to judgment, *Oviraptor* was falsely accused of stealing eggs that it may have been protecting or incubating (Fig. 24-1).

In addition to examining the evolution of the dinosaurs, in this chapter we will discuss aspects of dinosaur physiology and behavior that are revealed by the fossil evidence. But the Mesozoic world was inhabited by much more than dinosaurs. We will examine the other inhabitants of this wonderfully diverse landscape, populated by a host of other reptiles and nonreptiles.

Mesozoic Marine Life

The Permo–Triassic extinction changed the look of the marine realm tremendously, but lost diversity usually provides an opportunity for new adaptive radiation. Diversification during the Mesozoic is especially evident in the evolutionary records of several important invertebrate phyla and certain groups of fish. The Mesozoic oceans also witnessed important changes in predator/prey relationships with the appearance of several groups of marine reptiles, which quickly assumed positions at the top of the food chain.

Invertebrate Fauna

Many of the major invertebrate groups that dominated the Paleozoic oceans were missing as the Mesozoic Era began. Trilobites, rugose and tabulate corals, blastoids, and fusulinid foraminifera were gone forever, and other important groups that were severely reduced, such as brachiopods and crinoids, never recovered. The ecological roles of many of these organisms were assumed by new groups of corals, planktonic organisms, and especially by several classes of mollusks.

Mollusks The diversification of mollusks is one of the greatest success stories of Mesozoic evolution. The growing abundance of different bivalves in the benthic community allowed them to replace the severely reduced brachiopods, which had been so dominant in the Paleozoic benthic communities. Oysters, in particular, were successful in the Mesozoic and were quite large and abundant in the shallow ocean along the Atlantic and Gulf coastal plains (Fig. 24-2). **Rudists,** the most unusual bivalves of the Mesozoic, had shells that consisted of a large cone-shaped valve, up to 1 meter (3 feet) tall, that was attached to the sea floor, and a smaller valve that formed a lid (Fig. 24-3). Rudist shells grew together in great masses and by the Middle Cretaceous Period had replaced corals as the dominant reef-forming organisms in the shallow oceans.

The greatest Mesozoic mollusk diversification was among the cephalopods, two orders of which, the nautiloids and ammonoids, appeared during the Paleozoic. Of the ammonoids, one subgroup, the **ammonites,** underwent an evolutionary frenzy during the Jurassic and Cretaceous. The complex patterns of the sutures on the shell wall, formed by the partitions of the inner chambers, allowed differentiation of many genera of these cephalopods (Fig. 24-4). What evolutionary purpose was served by the intricate folding of the septa of ammonite shells? Some paleontologists suspect that

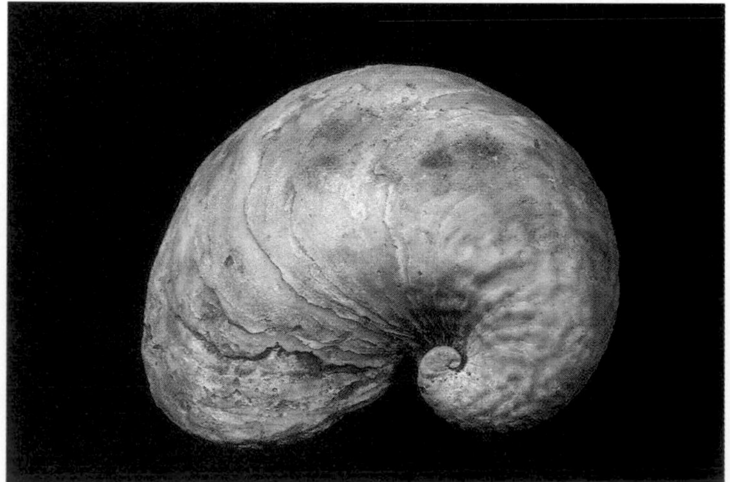

Figure 24-2 Oysters of the Jurassic. *Gryphia* (left) and *Exogyra* (above) were common bivalves in the Jurassic shallow marine environment.

Figure 24-3 Rudist bivalves. These unusually shaped bivalves grew in large clusters to form tropical reefs during the Cretaceous.

these complexly folded partitions were a means of reinforcing the shell against great water pressure, allowing the animals to inhabit a wider range of depths. Another possible function is that the increased surface area of the suture gave the animal greater control over buoyancy. Regardless, as rapidly evolving, free-swimming predators, ammonites are among the most important index fossils for the Jurassic and Cretaceous periods. Another cephalopod group that was abundant in Mesozoic oceans consisted of squidlike animals called **belemnites.** These creatures had a hard internal shell (or conch) that formed a cigar-shaped fossil common in Jurassic and Cretaceous marine rocks (Fig. 24-5).

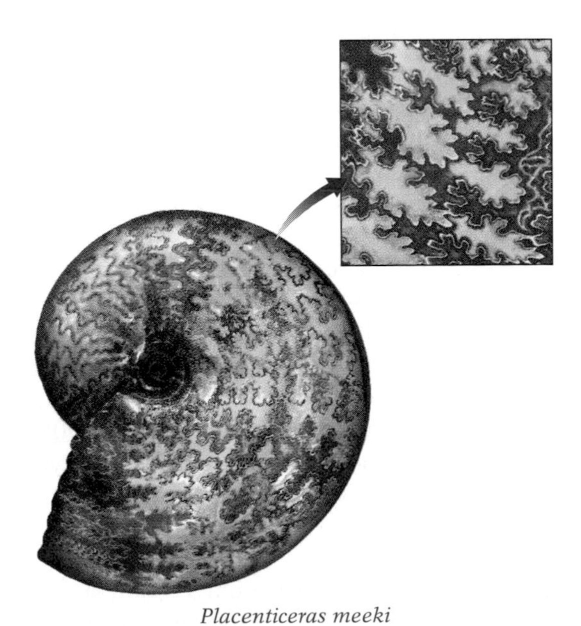

Figure 24-4 Ammonite suture patterns. The Cretaceous ammonite *Placenticeras* is recognized by the pattern of the sutures formed by the intricate folding of the septa.

Figure 24-5 Belemnites. Fossils of the belemnite conch are common in Jurassic and Cretaceous marine sediments.

Corals The familiar rugose corals of the Paleozoic oceans disappeared in the great Permo–Triassic extinction, but they were replaced eventually by the modern order of **scleractinian** corals, which appeared during the Middle Triassic. The fossil record doesn't reveal the evolutionary forerunners of this group, but soft-bodied anemones have been proposed as ancestors. Although their origin is unknown, the scleractinians had begun to form large coral reefs in the shallow oceans by the end of the Triassic, much as they do today. The success of these modern corals resulted from some key evolutionary advances. Noteworthy is that the skeletons of scleractinian corals are faster growing than most species with which they compete for space on the ocean floor, allowing them to crowd out other benthic organisms. This rapid growth is made possible in large part by a symbiotic relationship between the coral polyp and photosynthetic algae in reef-forming corals. The algae provide the coral with nutrients and oxygen but require that the coral grow in shallow, sunlit waters.

Echinoderms Although stalked echinoderms largely disappeared at the end of the Paleozoic, other echinoderms diversified and flourished in shallow Mesozoic seas. The number of varieties of echinoids (the group including sea urchins), starfish, and brittle stars expanded tremendously; all of these became abundant in the benthic habitat (Fig. 24-6).

Planktonic Organisms Important evolutionary developments also occurred at the base of the food chain. Foraminifera, which had been greatly reduced following the Permian

25

The Geology of the Cenozoic Era

Figure 25-1 Evidence of ancient glaciation. The parallel grooves in the bedrock of New York's Central Park were cut by glaciers that covered Manhattan during the last ice age.

Great geologic changes have indeed occurred in very recent history. Picture Manhattan, the heart of New York City, buried beneath 100 meters (328 feet) of ice while glaciers cover the New England states, the American Midwest, and all of Canada (Fig. 25-1). Also try to visualize the European capitals of Berlin, Warsaw, and Moscow covered by an ice sheet that extends over all of northern Europe and parts of Asia. This is what the world looked like at the peak of the last ice age, only 18,000 years ago; the effects of this glaciation have persisted even as the glaciers themselves have largely disappeared. A map of the United States drawn prior to this ice age would lack such familiar features as the Great Lakes and Cape Cod, because these (and many other features) were created entirely by the glaciers that buried the northern half of North America.

The great Swiss naturalist Louis Agassiz was not the first to recognize the moraines, erratic boulders, and scoured bedrock of his native Swiss Alps as the work of ancient glaciers. But in 1837, he was the first scientist to declare (with the utmost conviction) that these deposits were the result of a drastic climate change that caused a vast ice sheet to cover much of Europe and Asia. Agassiz, who eventually became one of the most famous professors in the history of Harvard University, was an evangelist for his "ice age" (the term is his); through careful presentation of the evidence, he convinced such luminary figures of nineteenth-century science as Charles Darwin and an initially skeptical Charles Lyell of his theory.

The Pleistocene Ice Age is only the most recent of the major geological events that occurred during the Cenozoic Era. Some of these were the continuation of events begun during the Mesozoic Era: the uplift of the Rocky Mountains, the building of the great mountain ranges of Europe, the final breakup of Gondwana, and the closing of the Tethys Seaway. All of these had their start during the Mesozoic. Significant changes in climate that occurred during the early part of the Cenozoic, resulting in part from these tectonic events, escalated through the late Cenozoic and culminated in the first major glaciation since the late Paleozoic. These same

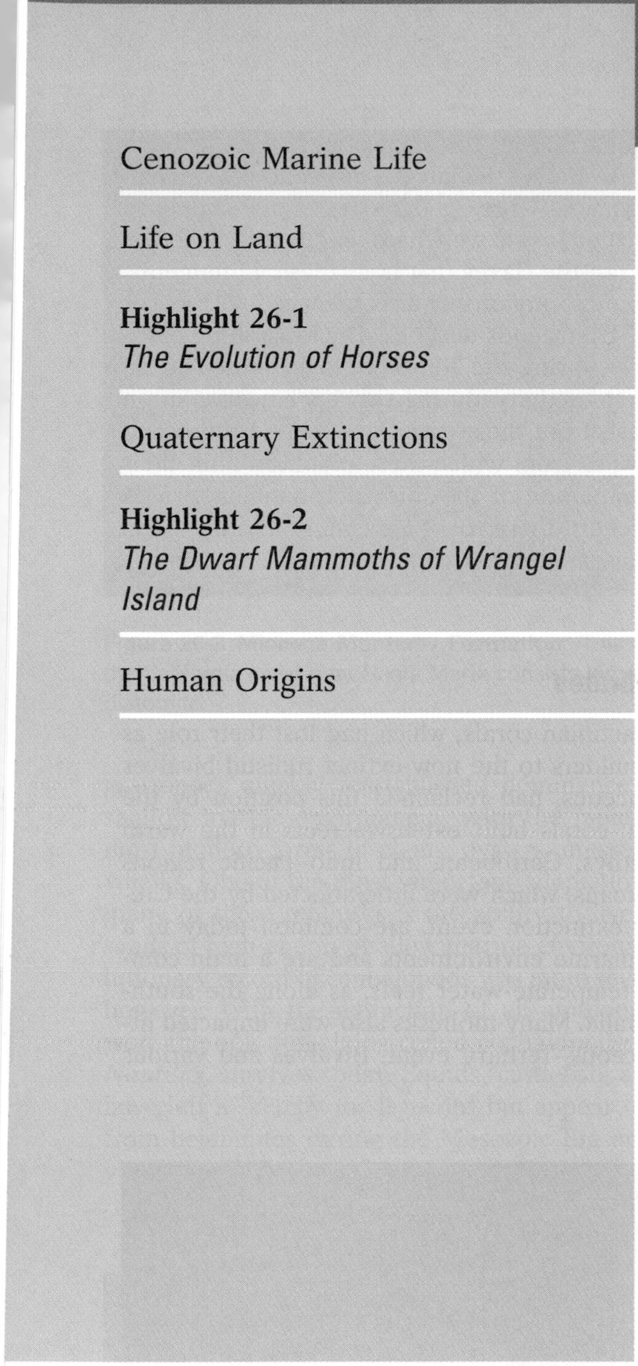

26

Cenozoic Life

Figure 26-1 Green River Formation. This fossil palm, preserved in Eocene lake sediments of the Green River Formation in southern Wyoming, attest to the tropical conditions that prevailed during the Eocene.

About 50 million years ago (during the Eocene Epoch), portions of Utah, Wyoming, and Colorado were covered by exceptionally large lakes. The fish that were abundant in these lakes included a wide variety of bony fish and even stingrays. A warm, nearly tropical climate prevailed, allowing palm trees and ferns to grow around the shores, also home to frogs, ducklike birds, insects, and even bats (Fig. 26-1). Around the lake margins lived now-extinct mammals that somewhat resembled modern squirrels and foxes, as well as extinct rhinoceroses and hoofed mammals.

Fossils of all of these creatures are preserved in the sediments of the Green River Formation, the deposits of these lakes providing us with a window on the diversity of life during the early part of the Cenozoic Era. Mudstones, limestones, and marls of this formation contain excellently preserved remains of many of the plants and animals that lived in and around the lakes. Physical separation of the lake waters into an upper layer stirred by the wind and a lower layer that lacked oxygen allowed preservation of organic remains that fell to the lake bottom in the fine-grained sediments. Many of the fish fossils occur in individual layers containing thousands of skeletons, clearly the result of some deadly change in lake conditions; overturn of the lake waters, replacing the well-oxygenated upper layer with the oxygen-poor bottom layer, could be to blame. Fossil Butte National Monument in southwestern Wyoming preserves an area in which these fossils are particularly abundant.

The Earth is today populated by plants and animals descended from the survivors of the mass extinction that defines the boundary between the Mesozoic and Cenozoic eras. Adaptive radiation of those survivors, which (importantly) included a number of mammal groups, quickly restocked ecosystems with a diverse array of new inhabitants. The history of this diversification is recounted in this chapter, with particular attention to the evolution of mammals.

Figure 26-10 Saber-toothed cat. *Smilodon* had bladelike teeth that were used for slashing prey.

not restricted to land. The third superfamily, the *pinnipeds,* including seals, sea lions, and walruses, are carnivores in which the limbs have been modified to flippers; they diverged from other groups during the Oligocene and adapted to an aquatic existence. The pinnipeds are unrelated to the other order of marine mammals, the cetaceans (whales and porpoises), discussed earlier.

Hoofed Mammals Nearly 200 living species of mammals are distinguished by having hoofed feet. This is a broad group, including such widely differing animals as horses, goats, sheep, camels, deer, pigs, and antelope, to which the general term *ungulates* is applied. Paleontologists often use this term in a broader sense to encompass also those mammal groups descended from hoof-bearing ancestors, such as elephants and possibly whales. Early ungulates were already common by the late Paleocene Epoch, and two distinct evolutionary pathways were established by the early Eocene (Fig. 26-11). One involved the radiation of the **odd-toed ungulates,** the group that today includes horses, rhinoceroses, and tapirs; the other pathway radiated to the more diverse group of **even-toed ungulates,** represented today by deer, giraffes, bison, pigs, hippopotamuses, and many more.

The odd-toed ungulates (formally called *perrisodactyls*) appear to be the older of the two orders of hoofed mammals. Today, this group only accounts for about 10% of all hoofed animals, and is represented by the remaining species of horses (see Highlight 26-1), rhinoceroses, and tapirs (16 species in all). Rhinoceroses have had a particularly interesting evolutionary history. Although the earliest Eocene members of this group were hornless and no larger than a German shepherd, radiation of this group produced some members during the Oligocene that stood as tall as modern

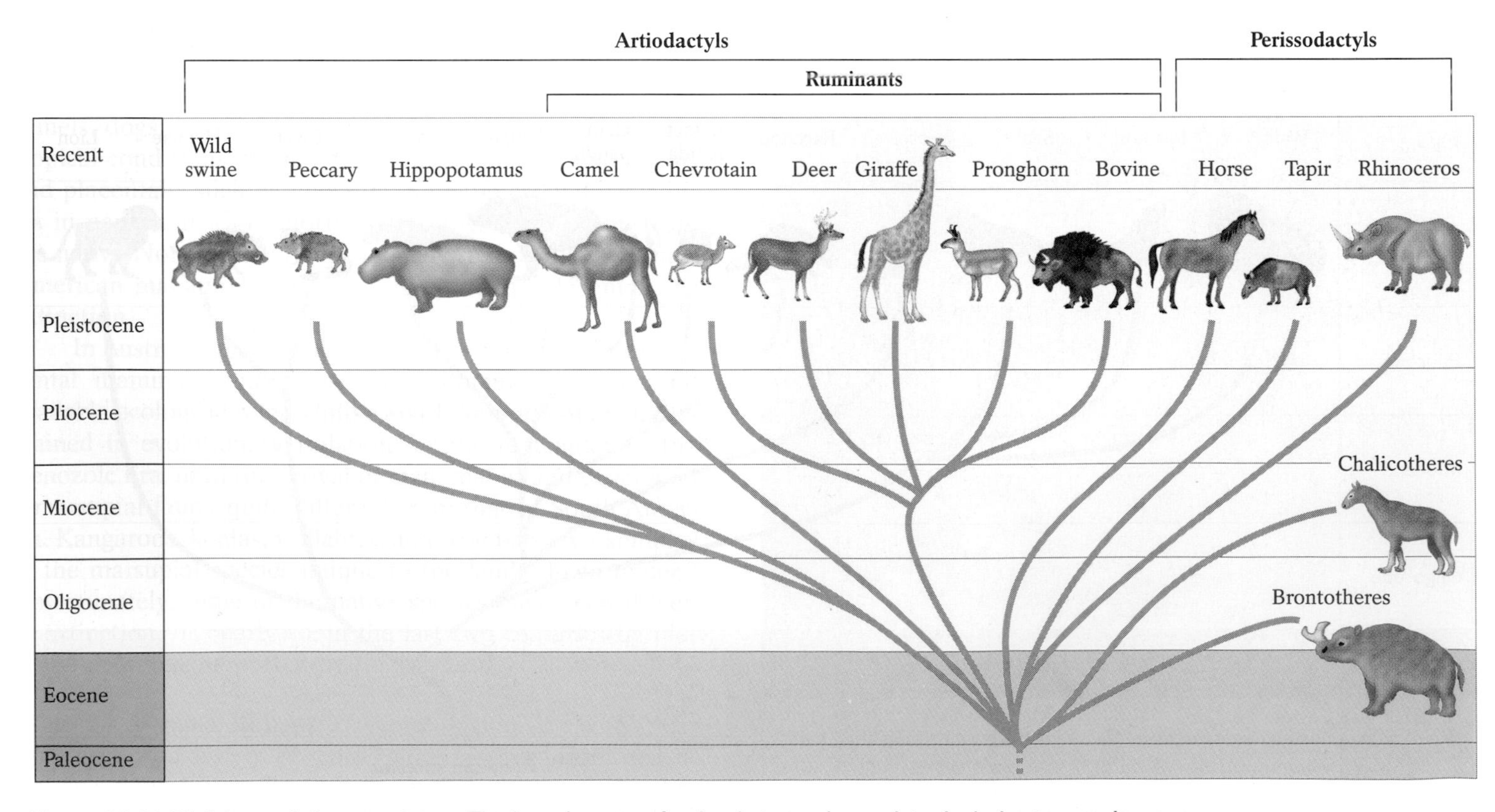

Figure 26-11 Divisions of the ungulates. The broad group of animals termed *ungulates* includes two main groups of hoofed animals: the odd-toed perrisodactyls and the even-toed artiodactyls.

Figure 26-12 *Brontotherium*. This giant beast stood almost 3 meters (10 feet) tall and lived during the Oligocene.

giraffes and others that lived hippolike in the water. Other extinct members of this order are equally fascinating. The *brontotheres* were enormous beasts that lived during the Eocene and earliest Oligocene epochs, distinguished by broad, forked horns over their snouts (Fig. 26-12). The odd-looking *chalicotheres* were large animals with horselike heads, relatively short hind limbs, long forelegs, and feet ending in claws rather than hooves (Fig. 26-13). These animals appeared during the Eocene and survived until the Pleistocene.

The earliest even-toed ungulates (formally known as *artiodactyls*) also appeared during the early Eocene Epoch

Figure 26-13 Chalicotheres. *Moropus*, a chalicothere that lived during the Miocene, was the size of a modern horse and had clawed feet.

and diversified rapidly. Although the number of species of odd-toed ungulates exceeded that of the even-toed ungulates during the Eocene Epoch, even-toed ungulates had gained the upper hand by the Oligocene and have continued to outnumber the odd-toed ungulates ever since, now accounting for some 170 species of living hoofed animals. The diversity of this group is surprising, including as it does sheep, camels, pigs, peccaries, giraffes, goats, deer, hippopotamuses, and many more families. The success of this order may be related to the development of the multichambered stomach, the chief characteristic of the **ruminants,** the largest group in this order, encompassing sheep, cattle, deer, antelope, goats, and others. More efficient digestion of food may have provided the advantage that allowed these even-toed ungulates to compete against their odd-toed cousins.

The oldest camels are known from the fossil record of the Eocene Epoch, but early camels were small and had four toes. This group diversified during the Miocene, by which time they had only two toes and had acquired forms resembling modern antelope and giraffes. You may be surprised to learn that most of the history of camel evolution occurred in North America. During the Pleistocene Epoch, the ancestors of modern camels migrated to the Old World across the Bering landbridge; subsequently, extinction eliminated this group in North America.

Members of the deer family first appeared during the Oligocene Epoch. Initially small and hornless, they soon diversified into forms with larger bodies and antlers. The most spectacular example of this radiation was *Megaloceros*, also called the "Irish elk," a Pleistocene deer that sported antlers measuring 3.5 meters (11.5 feet) across (Fig. 26-14; see page 560). The family of giraffes is thought to have split from the deer family during the Miocene, adapting to browsing strictly on leaves, although the characteristic long neck didn't evolve until much later.

The largest land animals alive today, the elephants evolved from hoofed ancestors and therefore are considered ungulates by some. The members of this family, which includes the well-known but extinct *mammoths* and *mastodons*, are naturally recognizable by their long trunks, an amazing modification of the nose, hence the family name, **proboscideans.** The earliest members of this family, which date to the Eocene Epoch, were trunkless, although they did possess tusks, another characteristic of this family. The trunk appears to be an adaptation to the increasing size of the body and head of these great beasts. Development of large, heavy heads required shortening of the neck for support, making it impossible to graze with the mouth; extreme elongation of the snout permitted easy gathering of food from the ground.

One branch of the evolutionary tree of proboscideans led to the *dinotheres*, an extinct group of Miocene and Pliocene elephants in which the tusks curved downward. Another branch of this tree led to the mastodons, mammoths,

Highlight 26-1 The Evolution of Horses

The fossil record of horses has long been recognized as powerful evidence for the theory of evolution; paleontologist O. C. Marsh (famous for his dinosaur discoveries) first described the lineage of horses in North America in the nineteenth century. From the first appearance of the oldest presumed ancestral horses during the late Paleocene Epoch to the emergence of the modern genus *Equus,* this line, with its many branches, has left a clear record of evolutionary change (Figure 1).

Horses belong to the order of perrisodactyls (the odd-toed ungulates), along with rhinoceroses, tapirs, and extinct forms such as the chalicotheres and brontotheres. The most recent evidence suggests that a small herbivorous mammal from the Paleocene of China, named *Radinskya,* is ancestral to this group, but this connection is still uncertain. The oldest generally accepted ancestor for the family of horses is the genus *Hyracotherium,* formerly known by the more elegant name *Eohippus* ("dawn horse"). This genus separated from other perrisodactyl ancestors during the late Paleocene, appearing first in Asia, then migrating to Europe and North America during the early Eocene. Although somewhat horselike in form, *Hyracotherium* was only about the size of a small dog, with four hoof-covered toes on the front feet and three on the hind feet. This may seem like a contradiction to its classification as a perrisodactyl, but only three of the toes on the front feet touched the ground.

Eocene horses, a lineage consisting of the genera *Orohippus, Epihippus,* and *Mesohippus,* exhibit more "horselike" characteristics, including increasing body size, longer skulls, and more complex teeth, but were still quite unlike modern horses. *Mesohippus* (Eocene to Oligocene) was still only about the size of a Labrador retriever and like its ancestors had low-crowned teeth; these were quite sufficient for browsing on the lush forest vegetation that grew before the global climate turned cold. This is in contrast to the high-crowned teeth of all later horses, an adaptation to the tough, abrasive grasses that spread across much of North America during the Miocene Epoch. Because these grasses contain bits of silica (which is abrasive), low-crowned teeth simply wear out faster and don't allow the animal to live very long. The early Miocene *Parahippus* appears to be the first genus to exhibit this adaptation.

Merychippus, a Miocene descendant of *Parahippus,* exhibits some important characteristics of modern horses. The forelegs ended in three toes, rather than four as in *Hyracotherium,* but more importantly, the center toe was greatly enlarged at the expense of the outer two. Gradually, these outer toes became vestiges, as in the late Miocene to Pliocene horse *Pliohippus,* before disappearing altogether. Combined with adaptations of the ankles, these new horses were more efficient at running across the open prairies that spread during the late Tertiary Period. Not all horses shared these adaptations, however. There were many branches of equine evolution that did not lead to the modern horse but instead became extinct; one of these, the hipparion group of horses, retained three equal-sized toes and lived primarily in mixed forest and open lands.

The modern genus of horse (*Equus*) appeared during the late Pliocene Epoch and radiated into a number of species, including zebras and asses, during the Pleistocene. These species moved freely between North America, where they had originated, and Asia; here they survived the extinctions that eliminated most large mammals at the end of the Pleistocene. Ironically, wild horses in this country are the descendants of domestic horses brought by Spanish explorers; native North American horses had died out thousands of years earlier.

Figure 1 "Family tree" of the horses. The path of evolution of the modern genus of horse, *Equus,* is long and has many branches.

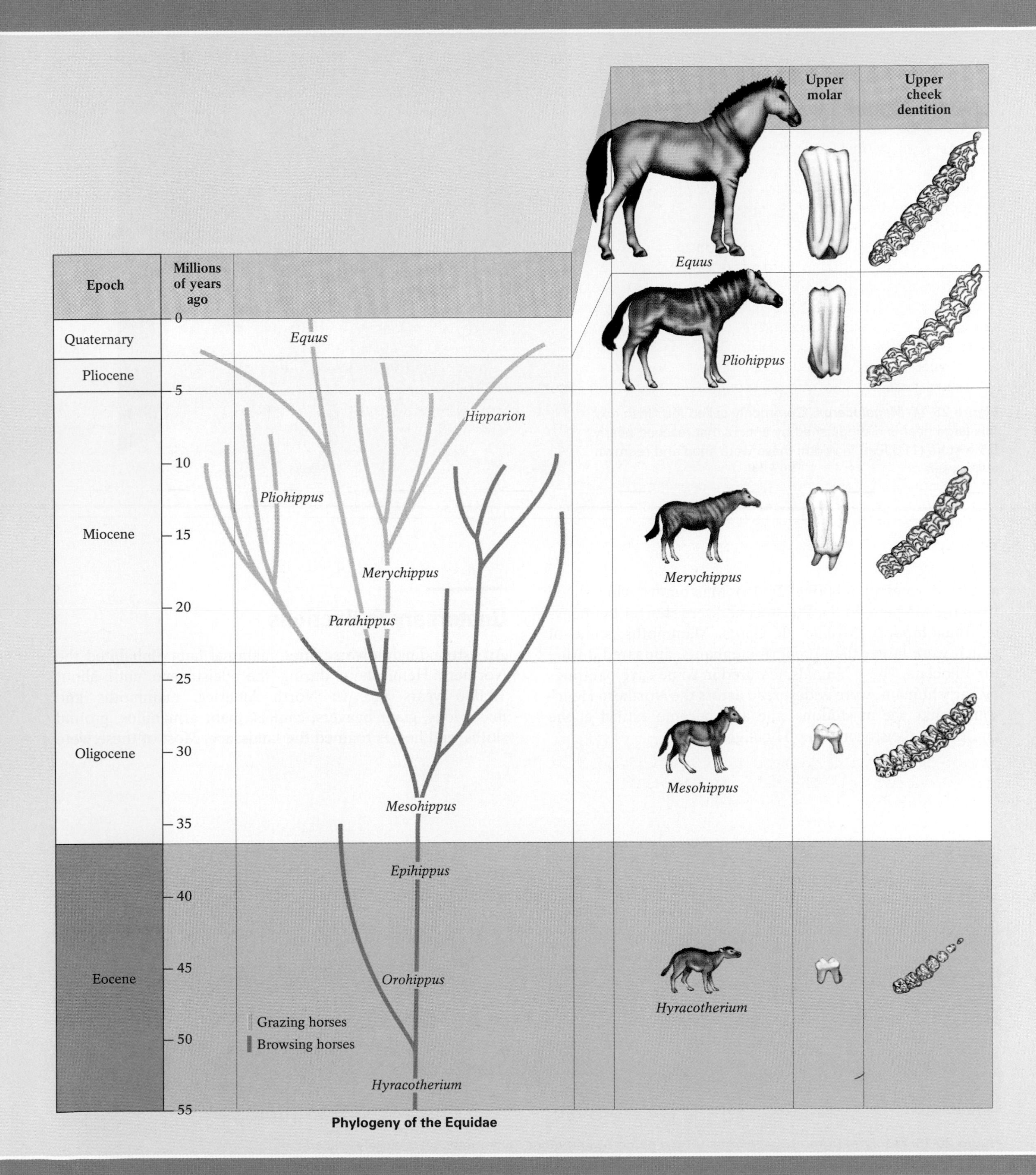

Phylogeny of the Equidae

Figure 26-14 ***Megaloceros*.** Commonly called the "Irish elk," this large deer is distinguished by antlers that reached nearly 3.5 meters (11.5 feet) in width; these were shed and regrown every year.

and modern elephants (Fig. 26-15). Mastodons, which lived from the Miocene to the Pleistocene, were shorter but heavier than modern African elephants. Mammoths, some of which were larger than modern elephants, appeared during the Pliocene. These animals, featured in some cave paintings by early humans, were widespread across the Northern Hemisphere, like the mastodons, and also became extinct at the end of the Pleistocene (see Highlight 26-2).

Quaternary Extinctions

An astoundingly diverse large mammal fauna inhabited the Northern Hemisphere during the Pleistocene until about 12,000 years ago. In North America, mammoths and mastodons, giant beavers, camels, giant armadillos, ground sloths, and horses roamed the landscape. Most of these were

Figure 26-15 Wooly mammoth. A member of the genus *Mammuthus,* mammoths were closely related to the modern Indian elephant. Adults stood up to 3.5 meters (11.5 feet) tall and weighed 6 to 8 tons. Several well-preserved specimens have been found frozen in the permafrost of Alaska.

Highlight 26-2 The Dwarf Mammoths of Wrangel Island

Generally speaking, mammoths, which once roamed freely from the Arctic to the Mediterranean and Gulf of Mexico, went extinct about 9500 years ago (at the end of the Pleistocene Epoch). But in the strictest sense, this isn't quite true. In 1991, mammoth teeth and bone fragments were discovered in the Arctic on Wrangel Island, a remote outpost almost 200 kilometers (124 miles) from the northeastern coast of Siberia. Radiocarbon dating of this material yielded ages for some of the mammoths of 13,000 to 20,000 years, nothing out of the ordinary. But scientists were astounded by the ages of other remains that prove that mammoths lived on this island until only about 3700 years ago. What's more, these recent mammoths were much smaller than the older mammoths, standing less than 2 meters (7 feet) high, compared with over 3 meters (10 feet) for the older mammoths.

How did mammoths survive on Wrangel Island long after they had become extinct elsewhere? Apparently, the mammoths first arrived here during the peak of the last glaciation when lower sea level allowed them to walk across a landbridge from the Asian continent. Rising sea level at the end of the Pleistocene drowned this bridge to the mainland, leaving a population isolated on the island. Fortunately for the stranded mammoths, Wrangel Island maintained a grassland habitat capable of supporting these large grazers, even as climate change eliminated most northern grasslands. But on an island with limited food resources, size is not an advantage. This ecological pressure, combined with a small gene pool, probably led to a gradual decrease in the size of the adults. Still, if this dwarf population successfully adapted to their environment, why did they ultimately succumb to extinction? Evidence exists for the arrival of Eskimo hunters by 3200 years ago, only 500 years later than the youngest mammoth remains. There is no proof, but it may be that these hunters wiped out the very last mammoths on Earth.

grazing animals, but carnivores were also plentiful; North American species of lion and cheetah, the sabertooth cats, and the dire wolf (larger than the modern wolf) preyed on the large herbivores. Between 12,000 and 8,000 years ago, however, all of these species disappeared. These late Quaternary extinctions are unlike other mass extinctions in Phanerozoic history in that the victims were limited to large land dwellers, sparing marine and smaller terrestrial taxa. Nearly 75% of genera of mammals weighing over 40 kilograms (88 pounds) in North America were lost, and 80% disappeared in South America. Europe and Asia, by contrast, suffered much lower rates of extinction (30% and 5% respectively). Most species of large, flightless birds present during the Pleistocene similarly met their demise.

Two competing and strongly debated hypotheses have been offered to explain the sudden loss of these animals. One suggests that many large animals were driven to extinction by a rapid climate change at the end of the Ice Age. We know that the Late Pleistocene climate was marked by very abrupt temperature changes; these changes caused disruptive shifts in the types and locations of vegetation. In some areas, increasing temperature and moisture caused forests to replace the steppe grasslands on which many of these animals depended; in other areas, decreasing rainfall caused landscapes to change from lush and vegetated to arid. In the face of this substantial habitat loss, many great beasts died out. This hypothesis is far from foolproof, however. Critics point out that abrupt climate changes had occurred during previous interglacial stages without sudden loss of these large species. In theory, many species should have been able to "outrun" extinction by migrating to new habitats as the old ones changed.

An alternative explanation, dubbed the **overkill hypothesis,** blames hunting by humans for the lost species. This notion is particularly attractive because human hunters, in contrast to animal predators, tend to prey on the largest game available. Evidence for hunting at the time of the large animal decline certainly exists; spear points have been found with the bones of giant bisons and mastodons. Also, the timing of these extinctions in the Americas coincides with the widespread appearance of people on these continents. This hypothesis is consistent with the lower extinction rates in Europe and Asia, where large animals and human hunters had long coexisted and the animals were presumably familiar with humans. By contrast, the animal populations in the Americas had no such experience with hunters when people first arrived and so may have been easier prey. Interestingly, most large animals had disappeared from Australia about 40,000 years ago, approximately the time at which humans first arrived on that continent. But this hypothesis is not without flaws, either. The major criticism is the lack of archaeological evidence for widespread killing. Tempting as it may be to visualize bands of hardy hunters slaughtering entire herds of mammoths, no mass bone beds that document mass killing have been found. Opponents of this hypothesis maintain that humans were simply too few and widely dispersed to have hunted so many species to extinction.

The ultimate cause of these extinctions may combine elements of both hypotheses. At a time of tremendous environmental stress, large animals may have existed in small populations, living precariously in marginal environments, and may have been particularly vulnerable to overhunting.

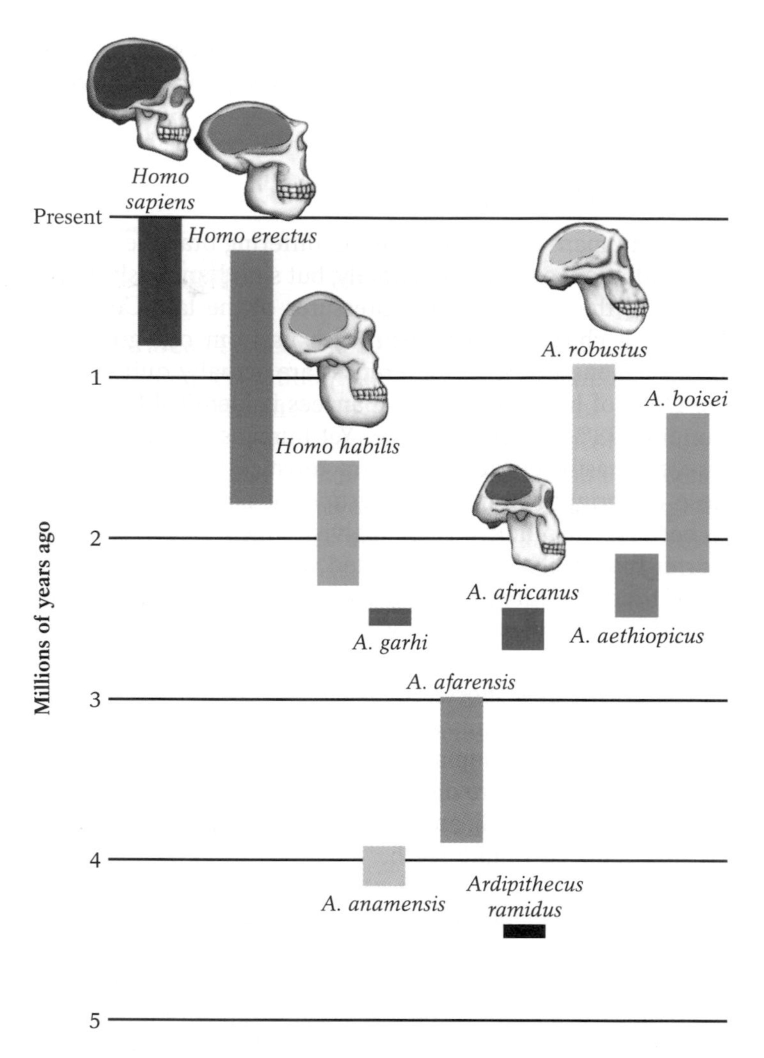

Figure 26-19 Human ancestors. Although the age ranges and identification of some species remain in question, the human "family tree" displays a surprising number of branches.

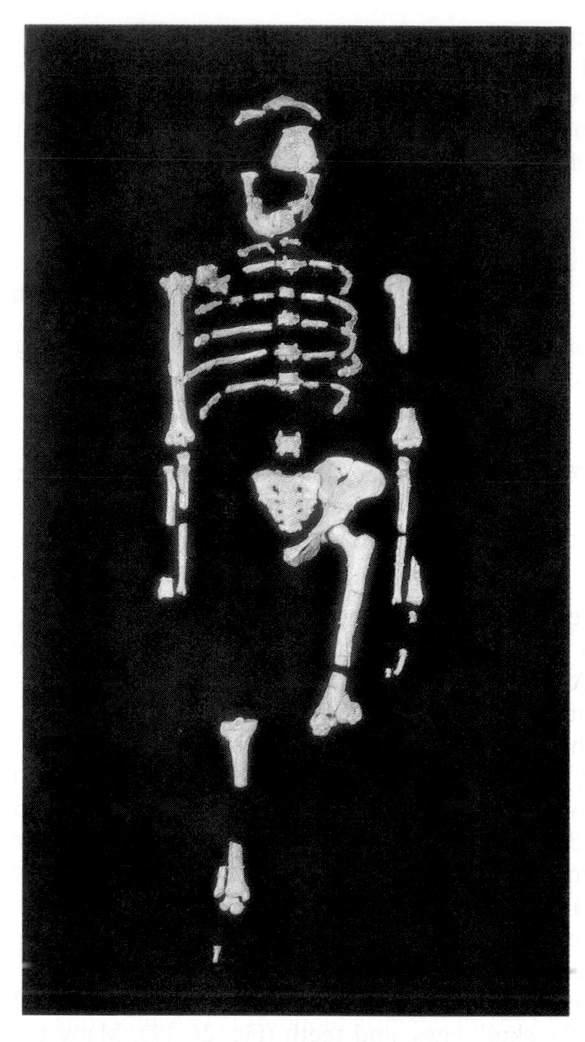

Figure 26-20 *Australopithecus afarensis.* This skeleton, named "Lucy," is the fossil of a female that stood about 1 meter (3 feet) high and lived about 3.5 million years ago.

leaving the fossil record on which we depend for this story. In northern Ethiopia, fossils of creatures from this time have been found that display a mix of human and apelike features. Named *Ardipithecus ramidus*, the fossils are tentatively considered the oldest member of the hominid family, although scientists are still uncertain that this creature actually walked upright.

Australopithecines We have no such reservations about slightly younger fossils, with a maximum age of 4.2 million years, assigned to the genus *Australopithecus.* Members of this genus, collectively termed **australopithecines,** were much smaller than modern humans; the males were only 1.3 meters (4 feet) tall on average, and the females about 1 meter (3 feet). Their brains were larger than those of modern chimpanzees, but still only about one-third the size of the brain of a modern human. Although australopithecines were apelike in some respects, particularly in regard to the face, which had a large, protruding jaw and bony ridges above the eyes, the pelvis design clearly indicates that they walked upright (Fig. 26-20). The most dramatic proof of this occurs in a tuff bed in Tanzania where several australopithecines walked in wet volcanic ash following an eruption 3.6 million years ago (Fig. 26-21). Despite this unambiguous proof of bipedalism, however, australopithecines had proportionately shorter legs and longer arms than modern humans. They were not adapted to walking great distances and still may have spent some of their time in trees; here they could gather food or escape predators.

Between their appearance and their extinction sometime around 1.5 million years ago, the australopithecines radiated into something between three and seven species, depending on which scientist you talk to. Some fossils that are

Figure 26-21 Standing erect. These footprints at Laetoli, Tanzania, were formed by a pair of australopithecines walking upright through wet volcanic ash following an eruption nearly 3.6 million years ago.

assigned to separate species on the basis of a single specimen may in fact belong to other described species. The oldest of these is *A. anamensis,* which lived between 4.2 and 3.9 million years ago. This early hominid is generally regarded as the ancestor of the more famous *A. afarensis* (3.9 million to 3.0 million years ago), made famous by the fossil of "Lucy," (see Fig. 26-20), named by its discoverer, Donald Johanson, after a song by the Beatles. These species, along with the younger *A. africanus* (3.0 million to 2.3 million years ago), form a group of smaller, more lightly built individuals with smaller teeth, referred to as the **gracile** australopithecines. They contrast with another group of larger, heavier-boned species, called the **robust** australopithecines, that lived later in the range of this genus (some researchers group these species in a separate genus, *Paranthropus*). The robust species include *A. boisei* (2.3 to 1.4 million years ago) and *A. robustus* (1.9 to 1.5 million years ago). You would be forgiven for assuming that our own genus, *Homo,* descended from the later, robust australopithecines, but most scientists believe differently. Instead, we think this branch of the family tree was an evolutionary dead end.

One of the most tantalizing recent discoveries is of a skull assigned the name *A. garhi,* dated at 2.5 million years. The facial features of this species are clearly those of an advanced australopithecine, but limb bones found nearby (which may not belong to the skull) indicate an individual much taller than other australopithecines. More importantly, stone tools and animal bones with cut marks found with the skull suggest that *A. garhi* made tools and used them to butcher animals for food, behavior that we associate with humans. Perhaps this is the path of human evolution.

Kenyanthropus The recent discovery of a 3.5-million-year-old hominid skull in Kenya threatens to overturn the current theories of human evolution. Anthropologist Meave Leakey has given the new fossil the name *Kenyanthropus platyops* ("flat-faced man of Kenya"). Although small-brained like its australopithecine contemporaries, the face of the newcomer was much less like that of a chimpanzee, with a flatter profile, less prominent brow ridges, smaller teeth, and larger cheekbones. In fact, *K. platyops* bears a strong facial resemblance to early members of genus *Homo,* who didn't appear until about 1.5 million years later. This similarity has fueled speculation that the australopithecines are not the true ancestors of modern humans after all, but that a separate limb of the family tree led to *H. sapiens.* At this time, however, it is still much too early to determine the evolutionary relationships among this new genus, the australopithecines, and the later genus *Homo.* Continued research is certain to reveal new discoveries that will shed light on this exciting area of evolutionary research.

The Genus *Homo*

The earliest representatives of our own genus had larger brains and smaller jaws and teeth than the australopithecines, presumably an adaptation by early humans to their improved ability to cut meat. Additionally, changes in the pelvic bones and longer leg bones would seem to indicate that these hominids spent all of their time on the ground, not in trees. These adaptations may have been provoked by climate change during the Late Pliocene, at which time glaciation in the Northern Hemisphere began. When East Africa became cooler and drier, much of the forest habitat was replaced by grassy savanna, and hominids could no longer depend on trees for safety and food. As a consequence, they were forced to travel greater distances across open ground in search of nourishment.

The fossil record of the genus *Homo* is every bit as fragmentary and confusing as it is for *Australopithecus,* and much debate surrounds the assignment of some fossils to separate species. The oldest known fossils of genus *Homo* date from 2.4 million years ago and may be of an early representative (or possibly an ancestor) of *H. habilis,* the "Handy Man," so called because it was long thought to be the first tool user. This species was no larger than the several species of australopithecines with whom it coexisted, but its brain was definitely larger, although still only about one-half the size of a modern human brain.

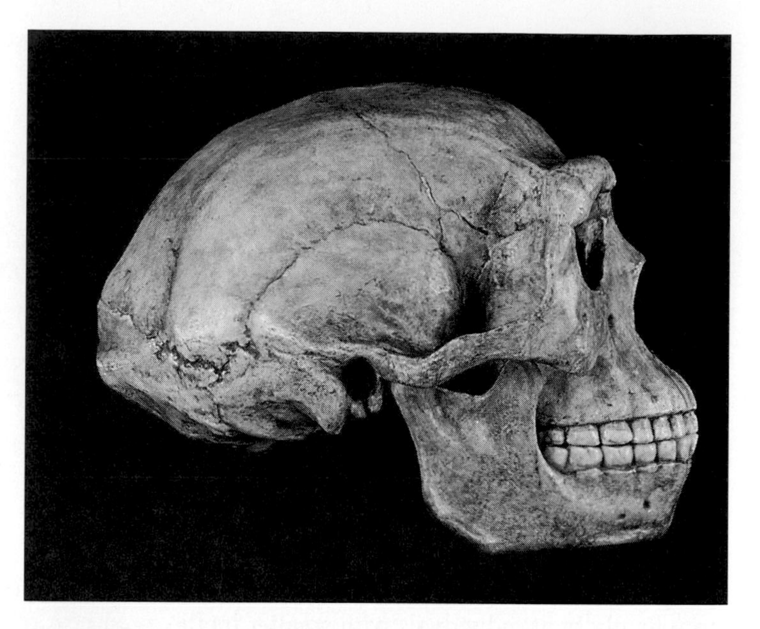

Figure 26-22 ***H. erectus.*** The brain of *Homo erectus* was significantly larger than that of its predecessors. *H. erectus* was the first hominid species to become widespread.

Homo erectus In contrast to *H. habilis,* whose fossils are unknown outside of Africa, ***Homo erectus*** (1.8 million to 250,000 years ago) was the first hominid to appear on other continents (Fig. 26-22). During the Early Pleistocene Epoch, these early humans made their way into Europe and Asia, migrating as far as China and Indonesia. What were these early wanderers like? In many respects, this species was much like modern humans, but only from the neck down. Although their bodies were comparable in size to modern humans, their thick-walled skulls still had considerably less brain capacity than the average modern person. The face was more primitive, too, with prominent brow ridges, a sloping forehead, and jutting jaw. Nevertheless, *H. erectus* exhibited major increases in brain and body size over *H. habilis.* This species also demonstrated unusually modern behavior, creating well-sculpted stone tools for use in hunting and butchering, and there is even evidence from some fossil sites that *H. erectus* used fire. These advances certainly help explain why this one species of human lasted considerably longer than any other.

Neanderthals Exactly what the next step was on our evolutionary trail isn't so clear. We know that between 200,000 and 30,000 years ago, Europe and the Near East were inhabited by people with larger, more muscular bodies than *H. erectus,* but with similar facial features (sloping forehead, prominent brow ridges, projecting jaw), and more importantly, larger brains (Fig. 26-23). Named **Neanderthal,** after the discovery of the first specimen in the Neander Valley of Germany (*tal* means "valley" in German), these sturdy people appeared in Europe well before the oldest known mod-

Neanderthal

Figure 26-23 Neanderthal. The Neanderthal people were heavily built, but slightly shorter than modern humans. Their skulls are distinguished by a protruding face, sloping forehead, and prominent brow ridges.

ern humans. But the origin, ultimate fate, and evolutionary relationship of the Neanderthals with modern humans is still cloudy and controversial.

Compared with modern humans, these people were slightly shorter on average, but more heavily built, and had brains that were slightly larger; in proportion to body mass, however, their brains were actually smaller. The Neanderthals are widely assumed to have evolved from *H. erectus* ancestors, but just when is not at all clear. More primitive forms of Neanderthals, dating back as far as 600,000 years, are sometimes assigned the name *H. heidelbergensis;* this group has been suggested as ancestral to the Neanderthals. Even older fossils (800,000 years ago) from Spain have been tentatively assigned the name *H. antecessor* and suggested as a common ancestor to lines leading to both modern humans and Neanderthals. This interpretation is based on preliminary research, however, and remains in doubt.

Because they were clearly more advanced than *H. erectus,* but physically distinct from modern humans, many researchers assign the Neanderthals to a separate species, *H. neanderthalensis.* Others disagree, classifying the group as a subspecies of our own (*H. sapiens neanderthalensis*). Comparison of the DNA sequences from several Neanderthal specimens and modern humans has provided very strong evidence of two genetically distinct species, incapable of interbreeding and producing fertile offspring. Differences in the DNA between Neanderthal specimens separated in age by thousands of years are substantially less than those between Neanderthals and modern humans. From these differences, scientists have estimated that the two species may have diverged 500,000 years ago.

Advocates of the subspecies interpretation offer as evidence several sets of fossil remains in which Neanderthal and *H. sapiens* features appear to be combined. In particular, controversy has erupted over a skeleton recently discovered in Portugal of a four-year-old child that seems to have a modern human face but a Neanderthal body. The presence of both modern and Neanderthal traits in a substantial number of individuals would indicate interbreeding between these groups; this would require that they were in fact members of the same species. However, these specimens are exceedingly rare, and their interpretation is controversial. Clearly, much more evidence is required to resolve this debate.

Contrary to the stereotype of brutish cave dwellers, Neanderthals developed a surprisingly sophisticated culture. Although it is true that they utilized caves for dwelling, they also sometimes constructed crude shelters, fashioned stone implements more carefully than *H. erectus,* and hunted game with spears. They evidently were skilled hunters; the chemistry of Neanderthal bones indicates that their diet consisted entirely of meat. Their skeletons sometimes display partially healed injuries, showing that injured members were cared for, and their dead were sometimes buried in ritualistic fashion. The Neanderthals in Europe endured harsh conditions during much of the Pleistocene Ice Age yet disappeared soon after the arrival of modern humans in Europe for reasons unknown.

H. sapiens The discussion thus far has yet to answer the fundamental question of how *our* species originated. In answer, two opposing theories have arisen, each offering its own supporting evidence. The older, traditional theory suggests a **multiregional origin** of *H. sapiens;* humans evolved from a common ancestral stock, possibly *H. erectus* (who had so successfully populated much of Eurasia); *H. sapiens* evolved from these populations simultaneously in several locations. Limited intermingling of these geographically isolated groups allowed them to maintain genetic similarity but permitted development of the regional differences we call races.

The newer theory, referred to as **out of Africa,** was hotly contested when first proposed in the 1980s but has gradually gained proponents. According to this concept, *H. sapiens* evolved in Africa sometime between 100,000 and 300,000 years ago, and a group of these modern people migrated out of Africa no more than 100,000 years ago to populate the rest of the world. Although the fossil evidence for this theory is sketchy, strong support comes from studies of the DNA in the mitochondria of different human populations. Mitochondrial DNA is selected for study because this material is inherited only from the mother, with no changes except by random mutation. By comparing the differences in mitochondrial DNA in different groups of modern humans, and assuming a constant rate of mutation, it is possible to calculate the time since these different populations diverged. Critics of this theory point out that it is not at all certain that the rate of mutation has remained constant, and therefore that this calculation is questionable.

The most recent studies of the rates of mutation of DNA in both mitochondria and in the Y chromosome seem to confirm the "out of Africa" theory but also suggest that patterns of migration were more complex. The latest thinking is that an initial group of anatomically modern humans left Africa about 100,000 years ago but were stopped in the Middle East, perhaps because the way was blocked by Neanderthals. The true migration didn't occur until 40,000 to 50,000 years ago, when a group of more culturally advanced people left Africa and initially spread east through Asia, only later circling back west into Europe.

Inhabiting the World These modern humans, distinguished from Neanderthals by taller, but lighter bodies, vertical foreheads, and prominent chins, appeared in Europe about 35,000 years ago, and the Neanderthals disappeared within several thousand years. Did the Neanderthals become extinct as a species because they couldn't effectively compete with technologically more competent newcomers, or did they lose their identity through interbreeding with them? These are questions that cannot be answered with certainty as of yet.

We know that the European newcomers were nomadic hunters with a distinctive culture, called **Cro-Magnon,** that survived until about 10,000 years ago. This culture was characterized by the manufacture of tools from stone, wood, and bone, as well as decorative arts, including jewelry of shells, teeth, and ivory, and clay and ivory sculpture. They fashioned musical instruments (bone flutes), practiced elaborate burial rituals for their dead, and left elegant paintings on the walls of caves in which they sometimes dwelled (Fig. 26-24). Some of these paintings, in Spain and southern France, feature scenes of hunting and of contemporary animals, including many large mammals now extinct.

Modern people spread over much of southern and Southeast Asia even before they migrated into Europe, settling in Indonesia and New Guinea and reaching Australia 35,000 to 40,000 years ago. Even during the time of lower sea level, this migration was an astounding feat, requiring the development of seafaring technology at a very early point in human history. The arrival of people in the Americas in particular has long been a particularly fiercely debated issue. The long-held view was that humans first migrated across the Bering landbridge near the end of the Pleistocene, around 12,000 years ago, when retreat of the ice sheet opened a land route for passage. This old theory has been attacked repeatedly by evidence, including implements and fire pits of possibly greater age (as great as 30,000 years) from scattered sites in North and South America. But how could these nomads have reached the New World before the retreat of the glaciers? We now think it possible that the seafaring technology that allowed settlement of Australia also could have permitted people to sail around the ice and reach the Americas long before the land route was ice free. If this was indeed the case, early settlement would have been concentrated in coastal areas now drowned by the rise in sea level. Exploration of the continental shelf in northwestern North America has found evidence that now-submerged coastal regions were once occupied by people, but proof that this occupation began before 11,000 years ago is still lacking. In any event, retreat of the glaciers from the Bering landbridge about 11,000 years ago provided an easy route for later migration, and many more people probably arrived at this time.

Figure 26-24 Cro-Magnon culture. Artwork by the Cro-Magnon, exemplified by the famous cave paintings in southern Europe, indicate a well-developed decorative culture.

With the worldwide arrival of *H. sapiens,* our history of human evolution is now finished. Does this mean that the evolution of our species is now complete? Thanks to technology, no part of humanity exists in reproductive isolation from the rest of the population. Therefore, it seems unlikely that a new human species could arise to compete with us. Technology has also freed many of us, for the most part, from the demands of natural selection; in our society, few of us need be concerned with eluding predators or obtaining enough food for survival, and most of us have a reasonably good chance to live long, disease-free lives and reproduce. At present, the greatest threat to our species would seem to be the damage we inflict on our environment (as our population grows) through pollution, exhaustion of natural resources, and (perhaps) human-induced climate change. To preserve our species, and many of the other species inhabiting our planet, we need to dedicate our brains (greatly enhanced by evolution) to solving these problems.

Chapter Summary

The diversity of marine life lost in the Cretaceous–Tertiary extinctions was quickly recovered during the Cenozoic Era. Single-celled organisms, especially foraminifera and diatoms, radiated during the early Tertiary Period. Scleractinian corals regained their position as the dominant reef-building organisms. Among mollusks, bivalves and gastropods evolved the diverse forms we see today and adapted to high-energy environments in particular, whereas cephalopods failed to recover from the K–T boundary extinctions. Successful vertebrates of the Cenozoic oceans include the **teleost** fishes, the diversification of which produced most modern fish, and sharks that grew to giant proportions. The largest marine predators, the whales, had evolved from carnivorous, hoofed ancestors that lived during the Paleocene Epoch to large, fully aquatic mammals by the Eocene.

On land, vertebrate evolution was greatly influenced by changes in plant species, in particular by the continued diversification of angiosperms. The appearance and spread of grasses in response to Tertiary climate change had significant consequences on the evolution of grazing mammals. The Cenozoic was also the time of the great radiation of bird species. Predatory birds, including a number of large, flightless varieties, appeared during the Late Tertiary Period.

Modern orders of mammals that appeared before the start of the Cenozoic include the **monotremes,** the egg-laying mammals, and the **marsupials,** the pouched mammals. During the Tertiary Period, marsupials evolved into very diverse forms in South America and Australia, where they were isolated from the evolution of **placental** mammals on other continents. Most South American marsupials became extinct during the Pliocene, when a landbridge to North America allowed migration of competing placental mammals. **Insectivores** appear to have been the earliest placental mammals, evolving during the Cretaceous Period. But they were followed by the appearance of many new orders during the Paleocene and Early Eocene. Among the new groups were the **edentates** (which included the extinct glyptodonts and giant ground sloths), the **rodents** (gnawing mammals), the **lagomorphs** (rabbits), the extremely diverse **carnivores,** and the hoofed mammals. This last group encompasses two orders, the **odd-toed ungulates,** today including rhinoceroses and horses, and the much more abundant **even-toed ungulates.** The success of the latter group, which includes sheep, cattle, camels, giraffes, and deer, is attributable in part to the efficient, multichambered stomach of the **ruminants.** One group of ungulates, the **proboscideans,** including modern elephants and the extinct mammoths and mastodons, evolved long trunks to help them graze. A wave of extinctions at the end of the Pleistocene Epoch eliminated many of the large mammals, particularly in the Americas. One hypothesis attributes the extinctions to abrupt climate shifts and the accompanying changes in vegetation. A competing idea, the **overkill hypothesis,** blames hunting by early man.

The evolutionary record of **primates,** the placental order that includes humans, can be traced as far back as the Paleocene Epoch. The two principal primate groups, the **prosimians** (today including lemurs, lorises, and tarsiers) and the **anthropoids** (to which monkeys, apes, and humans belong), may have diverged from a common ancestor by the Middle Eocene. The **hominid** family, to which we belong, appears to have descended from apelike creatures called **dryomorphs,** which lived during the Miocene Epoch.

A surprisingly large number of species have been found of the early hominid genus of **australopithecines,** which walked upright but were much smaller than modern humans and had much smaller brains. Although a larger group of **robust** australopithecines evolved later, the modern genus *Homo* apparently evolved from the lighter **gracile** australopithecines. Members of the genus *Homo* are distinguished by larger brains and by their manufacture and use of tools. ***Homo erectus*** was the first hominid species to migrate to different continents. Although the evolutionary record is somewhat cloudy, we know that sturdy, large-brained people called **Neanderthals** became widespread across Europe and the Near East during the Late Pleistocene. The origin of modern humans, *H. sapiens,* remains in dispute. Adherents of the theory of a **multiregional origin** believe that *H. sapiens* evolved simultaneously in several areas from ancestral populations. The newer **out of Africa** theory maintains that modern humans evolved in Africa and later migrated to the other continents. Modern humans appeared in Europe about 35,000 years ago with a distinct culture we call **Cro-Magnon.** People reached southeast Asia and Australia at about the same time, but the date of settlement of the Americas remains in dispute.

Key Terms

teleosts (p. 551)
monotremes (p. 553)
marsupials (p. 554)
placental (p. 554)
insectivores (p. 554)
edentates (p. 554)
rodents (p. 555)
lagomorphs (p. 555)
carnivores (p. 555)
odd-toed ungulates (p. 556)
even-toed ungulates (p. 556)
ruminants (p. 557)
proboscideans (p. 557)
overkill hypothesis (p. 561)
primates (p. 562)
prosimians (p. 562)
anthropoids (p. 562)
hominids (p. 562)
dryomorphs (p. 563)
australopithecines (p. 564)
gracile (p. 565)
robust (p. 565)
Homo erectus (p. 566)
Neanderthal (p. 566)
multiregional origin (p. 567)
out of Africa (p. 567)
Cro-Magnon (p. 568)

Questions for Review

1. What new environment did bivalves and gastropods exploit in the Cenozoic Era?

2. Why were cephalopods unable to recover the diversity they experienced during the Mesozoic Era?

3. What primitive characteristic distinguishes monotremes from all other mammals?

4. What is the probable cause of the extinction of the multituberculate mammals?

5. Speculate how the chalicotheres used their unusual adaptation to survive.

6. Name some modern mammal groups that evolved in North America yet are not native to this continent.

7. What is the evidence to support the overkill hypothesis for the large-mammal extinction at the end of the Pleistocene?

8. What trend in hominid evolution is most significant in terms of the success of later species?

9. If Neanderthals had larger brains, why were they unable to compete with modern humans?

10. What type of evidence is cited in support of the "out of Africa" theory for the origin of *H. sapiens?*

For Further Thought

1. Describe the role of tectonics in the history of marsupial evolution.

2. Why do the even-toed ungulates appear to have diversified much more successfully than the odd-toed ungulates?

3. Describe how tectonics and global climate changes might be related to the evolution of hominids.

4. Examine the overlapping ranges of the later australopithecine species and early members of genus *Homo.* Speculate how this overlap may have affected human evolution.

5. What type of fossil evidence would be required to confirm the "multiregional" theory for the origin of *H. sapiens?*

Appendix A

Conversion Factors for English and Metric Units

Length	1 centimeter	=	0.3937 inch
	1 inch	=	2.54 centimeters
	1 meter	=	3.2808 feet
	1 foot	=	0.3048 meter
	1 yard	=	0.9144 meter
	1 kilometer	=	0.6214 mile (statute)
	1 kilometer	=	3281 feet
	1 mile (statute)	=	1.6093 kilometer
Velocity	1 kilometer/hour	=	0.2778 meter/second
	1 mile/hour	=	0.4471 meter/second
Area	1 square centimeter	=	0.16 square inch
	1 square inch	=	6.45 square centimeters
	1 square meter	=	10.76 square feet
	1 square meter	=	1.20 square yard
	1 square foot	=	0.093 square meter
	1 square kilometer	=	0.386 square mile
	1 square mile	=	2.59 square kilometers
	1 acre (U.S.)	=	4840 square yards
Volume	1 cubic centimeter	=	0.06 cubic inch
	1 cubic inch	=	16.39 cubic centimeters
	1 cubic meter	=	35.31 cubic feet
	1 cubic foot	=	0.028 cubic meter
	1 cubic meter	=	1.31 cubic yard
	1 cubic yard	=	0.76 cubic meter
	1 liter	=	1000 cubic centimeters
	1 liter	=	1.06 quart (U.S. liquid)
	1 gallon (U.S. liquid)	=	3.79 liters
Mass	1 gram	=	0.035 ounce
	1 ounce	=	28.35 grams
	1 kilogram	=	2.205 pounds
	1 pound	=	0.45 kilogram
Pressure	1 kilogram/square centimeter	=	0.97 atmosphere
	1 kilogram/square centimeter	=	14.22 pounds/square inch
	1 kilogram/square centimeter	=	0.98 bar
	1 bar	=	0.99 atmosphere
Temperature	°F (degrees Fahrenheit)	=	°C(9/5) + 32
	°C (degrees Celsius)	=	(°F – 32)(5/9)

Appendix B

A Statistical Portrait of Planet Earth

Surface Areas

Landmasses	150,142,300 kilometers^2 (57,970,000 miles^2)
Oceans and Seas	362,032,000 kilometers^2 (138,781,000 miles^2)
Entire Earth	512,175,090 kilometers^2 (197,751,500 miles^2)

Distribution of Water, by Volume

Oceans and Seas	1.37×10^9 kilometers^3 (3.3×10^8 miles^3)
Glaciers	2.5×10^7 kilometers^3 (7×10^6 miles^3)
Groundwater	8.4×10^6 kilometers^3 (2×10^6 miles^3)
Lakes	1.25×10^5 kilometers^3 (3×10^4 miles^3)
Rivers	1.25×10^3 kilometers^3 (3×10^2 miles^3)

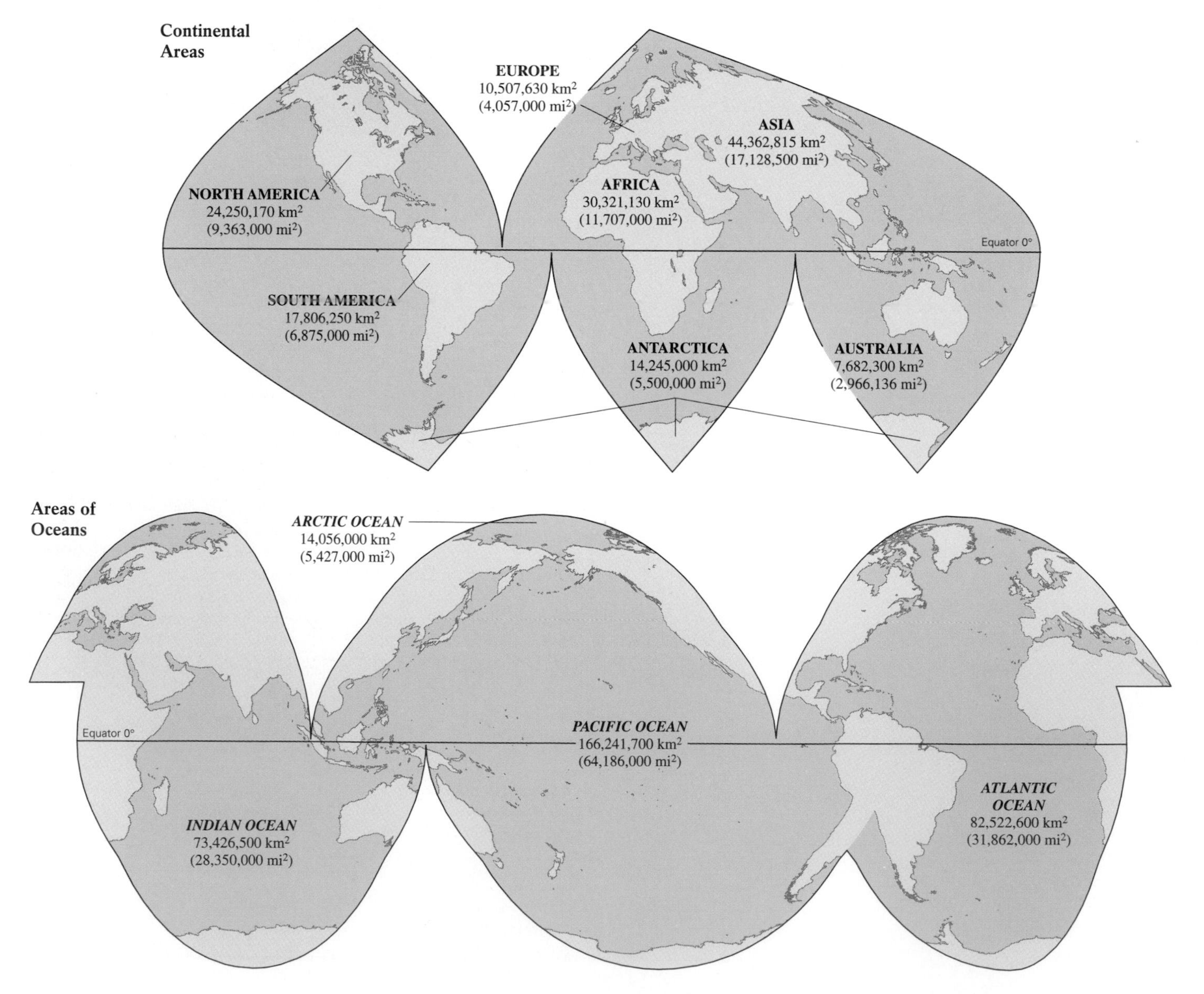

Elevations, Depths, and Distances

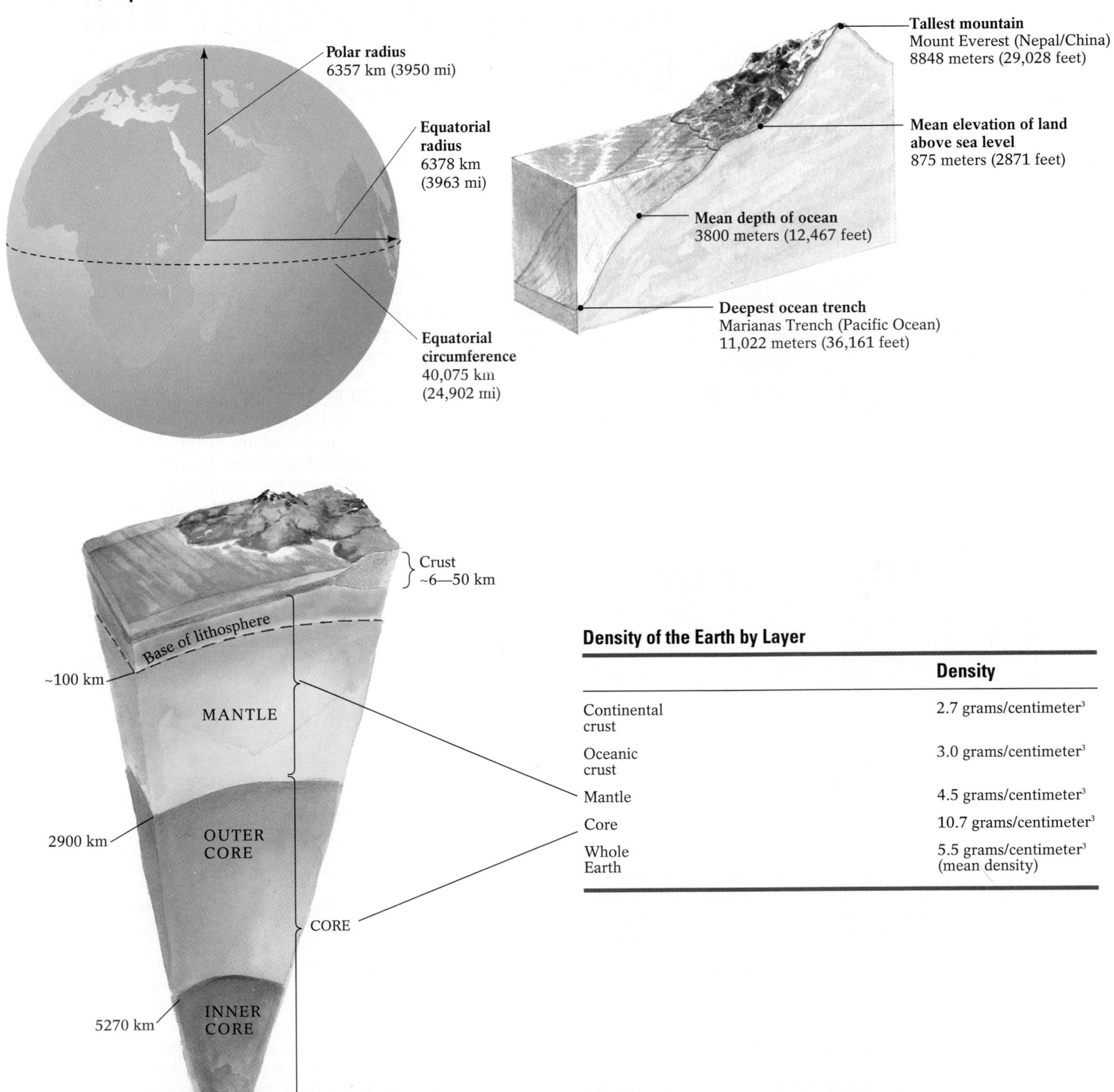

Density of the Earth by Layer

	Density
Continental crust	2.7 grams/centimeter^3
Oceanic crust	3.0 grams/centimeter^3
Mantle	4.5 grams/centimeter^3
Core	10.7 grams/centimeter^3
Whole Earth	5.5 grams/centimeter^3 (mean density)

Appendix C

Reading Topographic and Geologic Maps

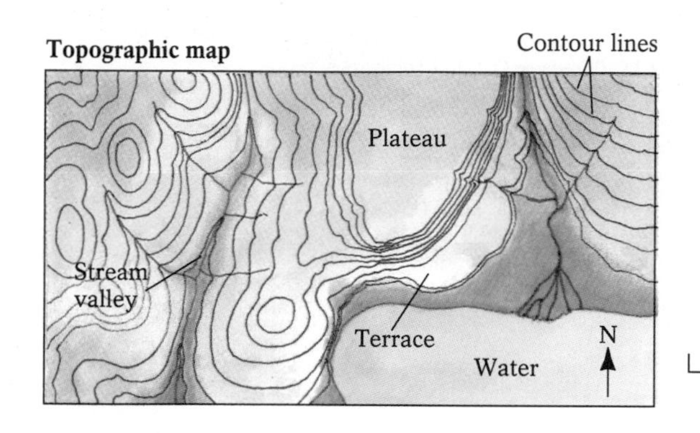

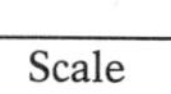

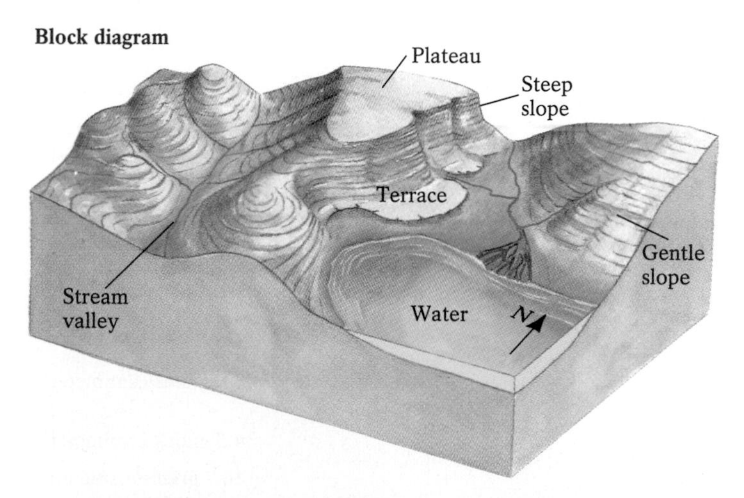

Throughout this textbook, maps have been used to show the locations of and relationships among various features. Geologists use maps for soil management, flood control, environmental planning, finding such resources as ores and groundwater, and determining optimal locations for fuel pipelines, highways, recreational areas, and the like.

Most maps show all or part of the Earth's surface, drawn to scale. A map's scale is usually shown at the bottom of the map. It may feature miles, kilometers, or any other convenient unit of measurement. Most maps from the United States Geological Survey are drawn to a scale of 1:24,000, meaning that 1 centimeter = 240 meters or 1 inch = 2000 feet; the latter would be read as "one inch on this map is equal to two thousand feet in real-world distance." The U.S. Geological Survey has been phasing in metric topographic maps using a scale of 1:100,000.

Most maps show two dimensions, marked by longitude (vertical) and latitude (horizontal) lines. A third dimension—elevation or depth—can be shown on *topographic maps,* which show relief using contour lines that mark specific heights above or depths below sea level. Every point on a contour line is at the same elevation. Every contour line closes upon itself—delineating the border of a discrete area—although its entire length may not be visible within the margins of a given map. A map's contour interval, or the distance between adjacent contour lines, is critical to both the usefulness and the appearance of the map: Contour lines that occur close together represent a steep slope; those that are farther apart represent a more gentle one. In an area of low relief, intervals designating elevation differences of only 5 or 10 feet are appropriate, whereas terrain with great relief may require intervals of 50 feet or more. On a given map, all contour intervals are the same.

In the past, accurate map-making depended predominantly upon actual measurements made on site by surveyors, geologists, and others. The advent of aerial photography, and then of space-based satellite photography, has made accurate standardized revisions possible.

The U.S. Geological Survey uses color to indicate specific features:

black and/or red solid lines : major roads

brown lines : contour lines

black : human-made structures, names

light red lines : town limits

blue : water features

green : wooded areas

white : open fields, deserts, and other nonvegetated areas

Dotted or dashed lines are used for temporary features (that is, solid blue represents a lake, whereas a dashed blue line marks a seasonal stream). Any special symbols used on a map are explained in its legend, which appears with the scale in the bottom margin.

A region's underlying geology can be represented on a *geologic map.* Symbols representing various types of rock have been standardized; such commonly accepted rock symbols have been used throughout this book. Standardized colors are used to show rock ages. The key to a geologic map's labeling, colors, and symbols usually appears at its side margins.

Columbus Quadrangle, New Jersey

Scale 1:24,000
Contour interval 10 feet

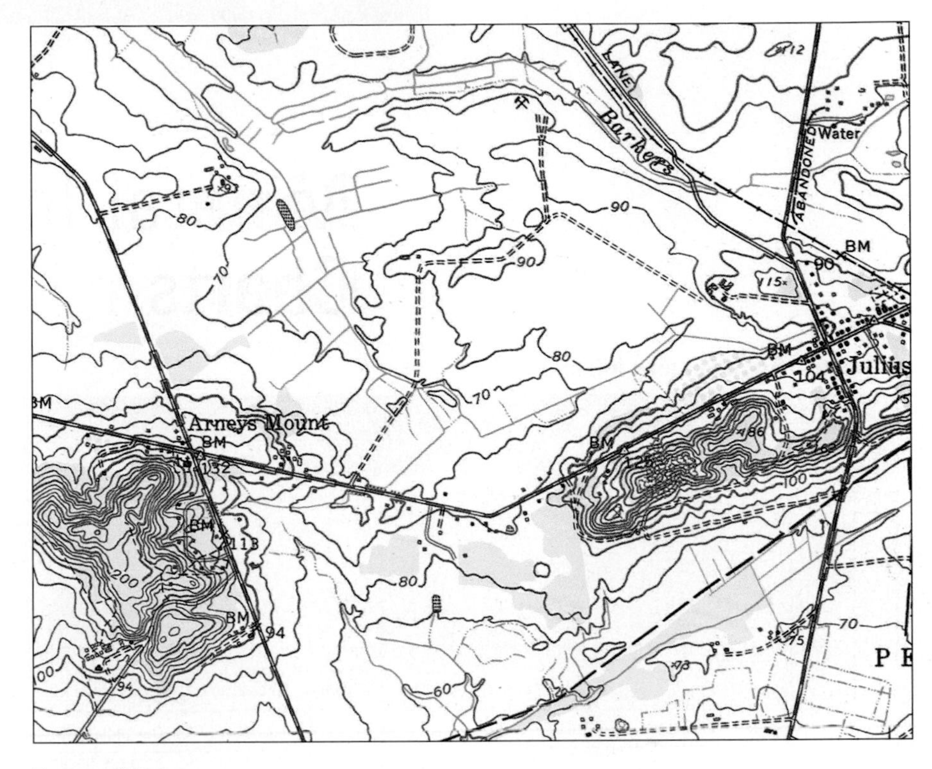

Topographic map

Geologic map

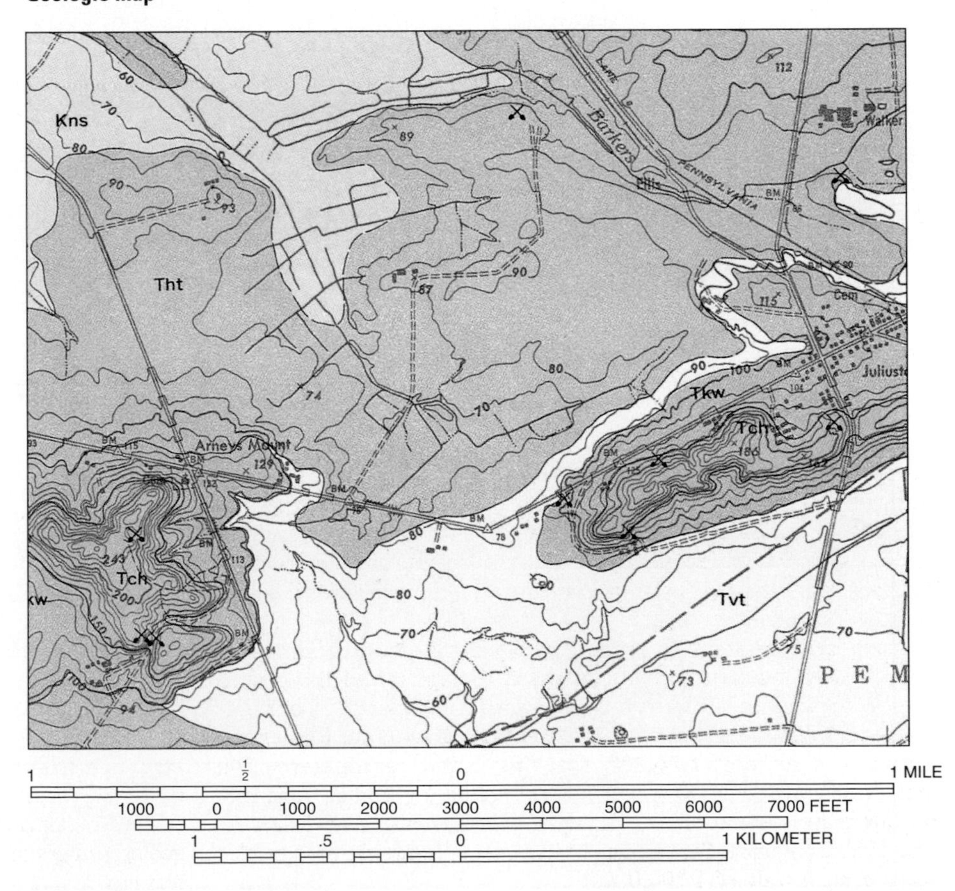

Tertiary Period Deposits

Tch Cohansey sand
Sand, quartz, light-gray to yellow-brown, medium- to coarse-grained, pebbly, ilmenitic, micaceous, stratified

Tkw Kirkwood formation
Sand, quartz, light-gray to tan, fine- to very fine-grained, clayey, micaceous, ilmenitic, kaolinitic, sparingly lignitic, massive-bedded

Tmq Manasquan formation
Sand, quartz, dark green-gray, medium- to coarse-grained, glauconitic, clayey

Tvt Vincentown formation
Upper member—calcarenite, quartz and glauconite, dusky-yellow to pale-olive, clayey; lower member—sand, quartz, dark-gray, poorly sorted, fine to coarse, clayey, glauconitic, entire formation very fossiliferous

Tht Hornerstown sand
Sand, glauconite, dusky-green, medium- to coarse-grained, clayey, massive-bedded

Cretaceous Period Deposits

Krb Red Bank sand
Lower member—sand, glauconite, dark grayish-black, coarse-grained, very clayey, micaceous, lignitic, massive-bedded (upper member not present)

Kns Navesink formation
Sand, glauconite, varying amounts of quartz, greenish-black to brown, medium- to coarse-grained, clayey

Kml Mount Laurel sand
Sand, quartz, reddish-brown to green-gray, poorly sorted, fine- to coarse-grained, glauconitic, massive-bedded

Appendix D

Mineral-Identification Charts

Most Common Rock-Forming Minerals

			MINERAL OR GROUP NAME	COMPOSITION/ VARIETIES	CRYSTAL FORM/OTHER DIAGNOSTIC FEATURES	CLEAVAGE/ FRACTURE	USUAL COLOR/ LUSTER	STREAK	HARDNESS
Light-colored; abundant in all rock types	SILICATES	Framework	Feldspar	Potassium (orthoclase) feldspar ($KAlSi_3O_8$)	Coarse crystals or fine grains	Good cleavage in two directions at 90°; cleavage surfaces not striated	White to gray or pink, with pearly luster	White	6
				Sodium, calcium (plagioclase) feldspar Albite: $NaAlSi_3O_8$ Anorthite: $Ca_2Al_2Si_3O_8$		Good cleavage in two directions at 90°; cleavage surfaces striated	White to gray, sometimes green or yellowish	White	
			Quartz	SiO_2	Six-sided crystals; individual or in masses	No cleavage; conchoidal fracture	Colorless or slightly smoky gray, pink or yellow	White	7
		Sheet	Mica	Muscovite $KAl_3Si_3O_{10}(OH)_2$	Thin, disc-shaped crystals	One perfect cleavage plane; splits into very thin sheets	Colorless or slightly gray, green, or brown, with vitreous luster	White	2–2½
				Biotite $K(Mg, Fe)_3AlSi_3O_{10}(OH)_2$	Irregular foliated masses	One perfect cleavage plane; splits into thin sheets	Black, brown, or green, with vitreous luster	White or gray	2½–3
				Chlorite $(Mg, Fe)_5(Al, Fe)_2Si_3O_{10}(OH)_8$			Yellowish, brown, green, or white	White or colorless	2–2½
Dark-colored; abundant in metamorphic and igneous rocks		Double Chain	Amphibole	Actinolite $Ca_2(Mg, Fe)_5Si_8O_{22}(OH)_2$	Long, six-sided crystals, fibrous or in aggregates	Two good cleavage planes at 56° and 124°	Pale to dark green or black, with vitreous luster. Pure actinolite white.	Pale green or white	5–6
				Hornblende $(Ca, Na)_{2-3}(Mg, Fe, Al)_5Si_6(Si, Al)_2O_{22}(OH)_2$					
		Single Chain	Pyroxene	Augite $Ca(Mg, Fe, Al)(Al, Si)O_6$	Short, four- or six-sided crystals	Two good cleavage planes at about 90°	Light to dark green, with vitreous luster	Pale green	5–6
				Diopside $CaMg(Si_2O_6)$			White to light green, with vitreous luster	White or pale green	
				Orthopyroxene $MgSiO_3$			Pale gray, green, brown, or yellow, with vitreous luster	White	
		Single Tetrahedra	Olivine	$(Mg, Fe)_2SiO_4$	Small grains and granular masses	No cleavage; conchoidal fracture	Grayish green or brown, with vitreous, glassy luster	White	6½–7
			Garnet	$(Ca_3, Mg_3, Fe_3, Al_2)n(SiO_4)_3$	12- or 24-sided crystals	No cleavage; conchoidal fracture	Deep red, with vitreous to resinous luster	White	6½–7½
Light-colored; abundant in sedimentary rocks	CARBONATES		Calcite	$CaCO_3$	Fine to coarsely crystalline. Effervesces rapidly in HCl.	Three oblique cleavage planes, forming rhombohedral cleavage pieces	White or gray, with pearly luster	White	3
			Dolomite	$CaMg(CO_3)_2$	Fine to coarsely crystalline. Effervesces slowly in HCl when powdered.		Colorless, white, or pink, and may be tinted by impurities, with pearly luster	White to pale gray	3½–4
	CLAY MINERALS	(Hydrous alumino-silicates)	Kaolinite	$Al_2Si_2O_5(OH)_4$	Very fine grains; found as bedded masses in soils and sedimentary rocks; earthy odor	Earthy fracture	White to buff, or tinted gray by impurities	White, off-white, or colorless	1½–2½
			Illite	$K_{0.8}Al_2(Si_{3.2}Al_{0.8})O_{10}(OH)_2$					
			Smectite	$Na_{0.3}Al_2(Si_{3.7}Al_{0.3})O_{10}(OH)_2$					

Accessory or Less-Abundant Rock-Forming Minerals

		MINERAL OR GROUP NAME	COMPOSITION/ VARIETIES	CRYSTAL FORM/OTHER DIAGNOSTIC FEATURES	CLEAVAGE/ FRACTURE	USUAL COLOR/ LUSTER	STREAK	HARDNESS
Light-colored; common in sedimentary rocks	SULFATES	Gypsum	$CaSO_4 \cdot 2H_2O$	Tabular crystals in fine-to-granular masses	One perfect cleavage plane, forming thin sheets; also two other good cleavage planes	Colorless to white, with vitreous luster	White	1–2½
	SULFATES	Anhydrite	$CaSO_4$	Granular masses	Three good to perfect cleavage planes at 90°	White, gray, or blue-gray, with pearly to vitreous luster	White	3–3½
		Halite	NaCl	Perfect cubic crystals, soluble in water. Tastes salty.	Three excellent cleavage planes at 90°	White or gray, with pearly luster	White	2½
Light-colored; mainly in metamorphic and igneous rocks	ALUMINO-SILICATES	Kyanite	Al_2SiO_5	Long, bladed, or tabular crystals	One perfect cleavage plane, parallel to length of crystals	White to light blue, with vitreous luster	White	5 along cleavage plane, 6½–7 across cleavage plane
	ALUMINO-SILICATES	Sillimanite		Long, slender crystals or fibrous masses		White to gray, with vitreous luster	White	6–7
	ALUMINO-SILICATES	Andalusite		Coarse, nearly square crystals	Irregular fracture	Red, reddish brown, or green, with vitreous luster	White	5–6
		Serpentine	$Mg_6Si_4O_{10}(OH)_8$	Fibrous or platy masses	Splintery fracture	Light to dark green or brownish yellow, with pearly luster	White	4–6
		Talc	$Mg_3Si_4O_{10}(OH)_2$	Foliated masses. Feels soapy.	Perfect in one direction, forming thin flakes	White to pale green, with pearly or greasy luster	White	1–1½
		Corundum	Al_2O_3	Short, six-sided crystals	Irregular fracture	Usually brown, pink, or blue, with adamantine luster	None	9
		Fluorite	CaF_2	Octahedral or cubic crystals	Cleaves easily	White, yellow, green, or purple, with vitreous luster	White	4
Dark-colored; common in metamorphic rocks	SILICATES	Epidote (paired tetrahedra)	$Ca_2(Al,Fe)Al_2Si_3O_{12}(OH)$	Usually granular masses; also slender prisms	One good cleavage direction, one poor	Yellow to dark green, with vitreous luster	White or gray	6–7
	SILICATES	Staurolite (single tetrahedra)	$Fe_2Al_9Si_4O_{22}(O,OH)_2$	Short crystals, some cross-shaped	One poor cleavage direction	Brown or reddish brown to black, with vitreous luster	Off-white to white	7
		Graphite	C	Scaley, foliated masses. Feels greasy.	One direction of cleavage	Steel gray to black, with vitreous or pearly luster	Gray or black	1–2
Dark-colored; common in all rock types		Apatite	$Ca_5(PO_4)_3(OH,F,Cl)$	Granular masses	Poor cleavage	Green, brown, or red, with adamantine or greasy luster	White	5
		Magnetite	Fe_3O_4	Granular masses	Uneven fracture	Black, with metallic luster	Black	5½
		Hematite	Fe_2O_3	Granular masses	Uneven fracture	Brown-red to black, with earthy, dull, to metallic luster	Brick red	5½
		Limonite	$2Fe_2O_3 \cdot 3H_2O$	Earthy masses	Uneven fracture	Yellowish brown to black	Brownish yellow	5–5½
Metallic luster; common in all rock types	SULFIDES	Pyrite	FeS_2	Cubic crystals or granular masses	Uneven fracture	Pale brass yellow, with metallic luster	Greenish black	6–6½
	SULFIDES	Galena	PbS	Cubic crystals or granular masses	Three perfect cleavage planes at 90°	Silver gray with metallic luster	Gray	2½
	SULFIDES	Sphalerite	ZnS	Granular masses	Six perfect cleavage planes at 60°	White to green, brown, or black, with resinous to submetallic luster	Reddish brown to yellow brown	3½–4
	SULFIDES	Chalcopyrite	$CuFeS_2$	Granular masses	Irregular fracture	Brass yellow	Greenish black	3½–4

Minerals and Elements of Industrial or Economic Importance

MINERAL OR GROUP NAME	COMPOSITION	USUAL COLOR/ LUSTER	STREAK	HARDNESS	OTHER PROPERTIES/ COMMENTS	ORIGIN
Asbestos (Chrysotile)	$Mg_3Si_2O_5(OH)_4$	White to pale green, with pearly luster	White	1–2½	Flexible, nonflammable fibers	A variety of serpentines; found mostly in metamorphic rock
Bauxite	$Al(OH)_3$	Reddish to brown, with dull luster	Pale reddish brown	1½–3½	Found in earthy, clay-like masses. Principal source of commercial aluminum.	Weathering of many rock types
Chalcopyrite	$CuFeS_2$	Yellow, with metallic luster	Greenish black	3½–4½	Uneven fracture; softer than pyrite. Iridescent tarnish. Most common ore of copper.	Hydrothermal veins and porphyry copper deposits
Chromite	$FeCr_2O_4$	Black, with metallic or submetallic luster	Dark brown	5½	Massive; granular; compact. Most common ore of chromium. Used in making steel.	Ultramafic igneous rocks
Copper (native)	Cu	Red, with metallic luster	Red	2½–3	Malleable and ductile. Tarnishes easily.	Mafic igneous rock (basaltic lavas); also in oxidized ore deposits
Galena	PbS	Silver-gray, with metallic luster	Gray	2½	Perfect cubic cleavage. Most important ore of lead; also commonly contains silver.	Hydrothermal veins
Gold (native)	Au	Yellow, with metallic luster	Yellow	2½–3	Malleable and ductile	Sedimentary placer deposits; hydrothermal veins
Hematite	Fe_2O_3	Brown-red to black, with earthy, dull, or metallic luster	Dark red	5½–6½	Granular, massive. Most important source of iron.	Found in all types of rocks. Commonly in contact-metamorphic aureoles around mafic igneous sills.
Magnetite	Fe_3O_4	Black, with metallic luster	Black	5½	Uneven fracture. An important source of iron. Strongly magnetic.	Found in all types of rocks
Platinum (native)	Pt	Steel-gray or silver-white, with metallic luster	Off-white, metallic	4–4½	Malleable and ductile. Occasionally magnetic.	Mafic igneous rocks and sedimentary placer deposits
Silver (native)	Ag	Silver-white, with metallic luster	Silver-white	2½–3	Malleable and ductile. Tarnishes to dull gray or black.	Mostly in hydrothermal veins; also in oxidized ore deposits
Sphalerite	ZnS	Shades of brown and red, with resinous to adamantine luster	Reddish brown	3½–4	Perfect cleavage in six directions at 120°. Most important ore of zinc.	Hydrothermal veins

Precious and Semi-precious Gems*

MINERAL OR GROUP NAME	COMPOSITION	USUAL COLOR/ LUSTER	STREAK	HARDNESS	OTHER PROPERTIES	ORIGIN
Beryl (aquamarine, emerald)	$Be_3Al_2(SiO_3)_6$	Blue, green, yellow, or pink, with vitreous luster	White	7½–8	Uneven fracture; hexagonal crystals	Cavities in granites and pegmatites, and schist
Corundum (ruby, sapphire)	Al_2O_3	Gray, red (ruby), blue (sapphire), with adamantine luster	None	9	Short, six-sided crystals; irregular, occasionally cleavage-like fracture ("parting")	Metamorphic rocks; some igneous rocks
Diamond	C	Colorless or with pale tints, with adamantine luster	None	10	Octahedral crystals	Peridotite; kimberlite; sedimentary placer deposits
(S) Garnet	$(Ca_3, Mg_3, Fe_3, Al_2)n(SiO_4)_3$	Deep red, with vitreous to resinous luster	White	6½–7½	No cleavage; 12- or 24-sided crystals	Contact-metamorphic and regionally metamorphosed rocks, and sedimentary placer deposits
(S) Jadeite (jade)	$NaAl(Si_2O_6)$	Green, with vitreous luster	White or pale green	6½–7	Compact fibrous aggregates	High-pressure metamorphic rocks
Olivine (peridot)	$(MgFe)_2SiO_4$	Light to dark green, with vitreous, glassy luster	White	6½–7	Uneven fracture, often in granular masses	Basalt, peridotite
(S) Opal	$SiO_2 \cdot nH_2O$	White with various other colors, with vitreous, pearly luster	White	5½–6½	Conchoidal fracture; amorphous; tinged with various colors in bands	Low-temperature hot springs; weathered near-surface deposits
(S) Quartz (includes amethyst, citrine, agate, onyx, bloodstone, jasper, etc.)	SiO_2	Colorless, white, or tinted by impurities, with luster depending on variety	White	7	No cleavage; six-sided crystals	Origin specific to gem variety. (Quartz found in all rock types except ultramafic igneous rocks.)
Topaz	$Al_2SiO_4(F, OH)_2$	Colorless, white, or pale pink or blue, with vitreous luster	White	8	Cleavage in one direction; conchoidal fracture	Pegmatite, granite, rhyolite
Tourmaline	$(Na, Ca)(Li, Mg, Al)$-$(Al, Fe, Mn)_6(BO_3)_3$-$(Si_6O_{18})(OH_4)$	Black, brown, green, or pink, with vitreous luster	White to gray	7–7½	Poor cleavage; uneven fracture; striated crystals	Metamorphic rocks; pegmatite; granite
(S) Turquoise	$CuAl_6(PO_4)_4$-$(OH)_8 \cdot 5H_2O$	Blue-green, with waxy luster	Blue-green or white	5–6	Massive	Hydrothermal veins
Zircon	$Zr(SiO_4)$	Colorless, gray, green, pink, or light blue, with adamantine luster	White	7½	Poor cleavage	Felsic igneous rocks and sedimentary placer deposits

*Gems are classified as semi-precious *(S)* if they are more accessible than precious gems and/or their properties are somewhat less valued.

Appendix E

How to Identify Some Common Rocks

Rocks are aggregates of minerals, and some are aggregates of organic materials. They are generally classified according to whether they are of igneous, sedimentary, or metamorphic origin, and then on the basis of their physical properties—primarily texture and mineral composition.

Igneous Rocks

Igneous rocks are formed when a magma cools and crystallizes. The chemical composition of a magma determines the minerals and rocks that will be formed when it cools. The rate at which a magma cools determines the size of the resulting crystals, which in turn determines the texture of the resulting rocks. Both composition and texture are used to classify igneous rocks.

Color is used only broadly as a diagnostic property for igneous rocks. The descriptions *light, dark,* and *intermediate* are generally agreed on. But many igneous rocks comprise two different-colored components; some blend various grays and pinks. Color, then, can help reduce the possible identities of a rock, but in most cases, color is not the defining property.

I. *Phaneritic*, or coarsely crystallized rocks: Composed of individual crystals of about 1 to 5 millimeters in length

A. If the rock contains quartz, it is probably **granite.**

B. If the rock contains no quartz, and from 30% to 60% feldspar, it is **diorite.**

C. If the rock contains no quartz and less than 30% feldspar, it is **gabbro.**

D. If the rock contains neither quartz nor feldspar, it is likely to contain ferromagnesian minerals and be ultramafic **peridotite** (consisting largely of olivine or pyroxene).

II. *Aphanitic* rocks: Individual crystals are fine-grained or microscopic

A. If quartz crystals can be identified in the rock, it is **rhyolite;** if the rock is light gray, white, or light green but too fine-grained to determine whether quartz is present, it is probably rhyolite.

B. If the rock contains no visible quartz crystals but does have approximately equal proportions of white or gray feldspar and ferromagnesian minerals, it is **andesite.**

C. If ferromagnesian minerals can be identified in the rock, it is **basalt;** if the rock is dark gray or black, and too fine-grained to identify any crystals, it is probably a basalt.

III. *Porphyritic* rocks: Contain macroscopic crystals (crystals that can be seen without the aid of a microscope) embedded in a matrix of smaller macroscopic or microscopic crystals. Use the descriptions above for aphanitic rocks to identify porphyritic rhyolites, basalts, and so on.

IV. *Glassy* rocks: Rocks look like solid glass. The most common is **obsidian,** which is rhyolitic in composition.

Sedimentary Rocks

Sedimentary rocks are made up of particles derived either from preexisting rocks or from organic debris. They are classified on the basis of their texture and mineral composition.

In identifying sedimentary rocks, the first step is to determine whether the rock contains carbonate minerals. This step is done by applying a drop of dilute hydrochloric acid to the rock surface. (*Note:* HCl must be used under close supervision to avoid burning one's skin or clothing.)

I. Rocks that do not effervesce (fizz), even when powdered—but may effervesce in some places, such as the fine cement between grains. Such rocks contain little or no carbonate minerals.

A. Rocks with **clastic** texture: Composed of grains in a cement matrix

1. If the grains are more than 2 millimeters in diameter . . .
   a. and are angular, the rock is **sedimentary breccia.**
   b. and are rounded, the rock is **conglomerate.**
2. If the grains are from 1/16 to 2 millimeters in diameter, and the rock feels gritty, it is **sandstone.**
   a. If more than 90% of the grains are quartz, the rock is **quartz sandstone.**
   b. If more than 25% of the grains are feldspar, the rock is **arkose.**
   c. If more than 25% of the grains are fine fragments of shale, slate, basalt, and the like, the rock is **lithic sandstone.**
   d. If more than 15% of the rock is fine-grained matrix material, the rock is **graywacke.**
3. If the rock is fine-grained, feels smooth, . . .
   a. and the grains are visible with a hand lens, the rock is **siltstone.**
   b. and the grains are invisible even with a hand lens, and the rock is . . .
      i. laminated, or layered, it is **shale.**
      ii. unlayered, it is **mudstone.**

B. Rocks with **crystalline** texture: Composed of microscopic, interlocking crystals
   1. If the rock dissolves in water, it is **rock salt.**
   2. If its crystals are fine to coarse, and have a hardness of 2 on the Mohs scale, the rock is **gypsum.**

C. Rocks with indeterminate texture
   1. If the rock is smooth, very fine-grained, and fractures conchoidally, it is **chert;** if it is dark in color, it is **flint.**
   2. If the rock is black or dark brown and breaks easily, soiling the fingers, it is **coal.**

II. Rock that effervesces strongly is **limestone.** Limestone may be clastic or crystalline in texture.

A. If the rock is clastic in texture and contains fossils, it is **bioclastic limestone.**
   1. If it is composed of whole, recognizable fossils, it is **coquina.**
   2. If it is very fine-grained, light-colored, and powdery, it is **chalk.**

B. If the rock is clastic and composed of small spheres, it is **oolitic limestone.**

C. If the rock is coarsely crystalline and contains different-colored layers, it is **travertine.**

III. Rock that effervesces only when hammered to a powder is **dolostone.** Dolostone may be crystalline, or less commonly, clastic in texture.

Metamorphic Rocks

Metamorphic rocks are rocks that have had their composition or structure changed by intense heat or pressure. Two factors determine the nature of a metamorphic rock: the composition of the parent rocks and the combination of metamorphic factors that acted upon them. Different circumstances can produce different textures; thus texture is the primary basis for metamorphic rock identification.

I. Is the rock *foliated*? If so, identify the type of foliation and, if possible, the rock's mineral content.

A. If the rock is fine-grained and readily splits into sheets, it is **slate,** and will have an earthy luster.

B. If the rock is slate-like but has a silken luster, it is **phyllite.**

C. If the rock contains flat or needle-like mineral crystals that are virtually parallel to one another, it is **schist.** A schist that primarily contains mica is a mica schist; there are, similarly, garnet mica schists, hornblende schists, and talc schists. A schist that contains serpentine is **serpentinite.**

D. If the rock consists of separate layers of light and dark minerals, it is **gneiss.** Light-colored layers consist of feldspars and perhaps quartz; dark-colored layers are probably biotite, amphibole, or pyroxene.

II. Is the rock nonfoliated? If so . . .

A. Is quartz its primary constituent? If so, the quartz grains will be interlocking and the rock will be hard enough to scratch glass. The rock is **quartzite.**

B. Does it consist of interlocking, coarse crystals of calcite or dolomite? If so, it is **marble.**

C. Does it consist primarily of dark grains too fine to be seen unaided? If so, it is probably **hornfels;** it may also contain a few larger crystals of less common minerals.

Appendix F

Classification of Living Things

This table presents an abbreviated listing of organisms prominent in evolutionary history. The traditional Linnaean classification system is used to organize this list, but as our knowledge of the history and evolutionary relationships among many of these groups is incomplete, this classification may differ from that in other texts. Similarly, the geologic ranges for many of these groups is imperfectly known.

DOMAIN BACTERIA

(also called Eubacteria)

Diverse prokaryotes including cyanobacteria, methane-producing bacteria, and disease-causing bacteria (Archean to Recent).

DOMAIN ARCHAEA (also called Archaeobacteria)

Prokaryotes genetically and chemically distinct from bacteria (Archean to Recent)

DOMAIN EUKARYA

All organisms with nucleated cells

KINGDOM PROTOCTISTA

Solitary or colonial unicellular organisms including:

PHYLUM CHLOROPHYTA

Green algae (Proterozoic to Recent)

PHYLUM CHRYOSOPHYTA

Coccolithophorids, golden-brown algae, and diatoms (Triassic to Recent)

PHYLUM SARCODINA

Protozoa with pseudopods for locomotion, including Foraminifera and Radiolaria (Cambrian to Recent)

KINGDOM PLANTAE

Photosynthetic eukaryotes including:

DIVISION BRYOPHYTA

Liverworts and mosses (Devonian to Recent)

DIVISION LYCOPODOPHYTA

Modern club mosses and Paleozoic scale trees, or lycopsids (Devonian to Recent)

DIVISION MAGNOLIOPHYTA

"Angiosperms," or flowering plants (Cretaceous to Recent)

DIVISION PSILOPHYTA

Primitive vascular plants lacking roots and leaves (Silurian to Recent)

DIVISION PTERIDOPHYTA

Ferns (Devonian to Recent)

DIVISION PINOPHYTA

"Gymnosperms," a diverse group of nonflowering, seed-producing plants including conifers, cycads, gingkoes, and seed ferns (Devonian to Recent)

DIVISION SPHENOPHYTA

Modern horsetails and scouring rushes as well as Paleozoic sphenopsids (Devonian to Recent)

KINGDOM FUNGI

Mushrooms, bread molds, slime molds, and other fungi (Proterozoic to Recent)

KINGDOM ANIMALIA

PHYLUM ANNELIDA

Segmented worms (Proterozoic to Recent)

PHYLUM ARTHROPODA

Subphylum Trilobita	Trilobites (Cambrian to Permian)
Subphylum Chelicerata	Arthropods with two main body sections, includes eurypterids, horseshoe crabs, and spiders (Cambrian to Recent)
Subphylum Crustacea	Crustaceans, including barnacles, crabs, lobsters, and ostracodes (Cambrian to Recent)
Subphylum Unirama	Arthropods bearing unbranched appendages, including velvet worms, centipedes, and insects (Cambrian to Recent)

PHYLUM BRACHIOPODA

Class Inarticulata	Brachiopods with valves of chitin or phosphate, lacking a hinge (Cambrian to Recent)
Class Articulata	Calcareous brachiopods with hinged valves (Cambrian to Recent)

PHYLUM BRYOZOA

Colonial filter-feeding organisms possessing a lophophore (Ordovician to Recent)

PHYLUM CNIDARIA

Class Scyphozoa — Most jellyfish (Proterozoic to Recent)

Class Anthozoa

- Order Tabulata — Tabulate corals, exclusively colonial (Ordovician to Permian)
- Order Rugosa — Rugose corals, solitary and colonial (Ordovician to Permian)
- Order Scleractinia — Solitary and colonial corals, including modern corals (Triassic to Recent)

PHYLUM ECHINODERMATA

Subphylum Crinozoa

- Class Blastoidea — Blastoids (Ordovician to Permian)
- Class Crinoidea — Ancient and modern crinoids (Ordovician to Recent)
- Class Cystoidea — Cystoids (Ordovician to Devonian)

Subphylum Asterozoa

- Class Asteroidea — Starfishes (Ordovician to Recent)
- Class Ophiuroidea — Brittle stars (Ordovician to Recent)

Subphylum Echinozoa

- Class Echinoidea — Sand dollars and sea urchins (Ordovician to Recent)
- Class Holothuroidea — Sea cucumbers (Ordovician to Recent)

PHYLUM MOLLUSCA

- Class Bivalvia — (also called Pelecypoda) Mollusks with bivalved shell, includes clams, mussels, oysters, and scallops (Cambrian to Recent)
- Class Gastropoda — Crawling mollusks with single chambered shell, usually spiraled, including snails and whelks, or no shell, such as slugs (Cambrian to Recent)

PHYLUM PORIFERA

Calcareous and siliceous sponges, possibly including stromatoporoids and archaeocyathids (Cambrian to Recent)

PHYLUM HEMICHORDATA

Animals possessing a notochord at some time during life; includes ancient graptolites and modern acorn worms (Cambrian to Recent)

PHYLUM CHORDATA

Subphylum Urochordata — Sea squirts and tunicates (range unknown)

Subphylum Vertebrata

- Class Agnatha — Jawless fish, includes extinct ostracoderms and modern lampreys (Cambrian to Recent)
- Class Acanthodii — Primitive spiny fish with jaws (Silurian to Permian)
- Class Placodermi — Primitive jawed fish, many with armored bodies (Silurian to Recent)
- Class Chondrichthyes — Cartilaginous fish including sharks, rays, and skates (Devonian to Recent)
- Class Osteichthyes — Bony fish
  - *Subclass Actinopterygii* — Ray-finned bony fish (Devonian to Recent)
  - *Subclass Sarcopterygii*
    - Order Crossopterygii — Includes rhipidistians (ancestral to amphibians) and coelacanths (Devonian to Recent)
    - Order Dipnoi — Ancient and modern lungfish (Devonian to Recent)
- Class Amphibia — Includes labyrinthodonts (Devonian to Triassic) and modern amphibians
- Class Reptilia
  - *Subclass Anapsida* — Reptiles in which the skull lacks temporal openings
    - Order Captorhinida — The earliest reptiles (Pennsylvanian to Permian)
    - Order Chelonia — Turtles (Triassic to Recent)
    - Order Mesosauria — Early aquatic reptiles (Pennsylvanian to Permian)
  - *Subclass Diapsida* — Reptiles with upper and lower temporal openings in skull
    - *Infraclass Lepidosauria* — Primitive diapsids, including snakes, lizards, and mosasaurs (Permian to Recent)
    - *Infraclass Archosauria* — Advanced diapsids having a skull opening in front of the eye socket
      - Order Thecodontia — Thecodonts (Permian to Triassic)
      - Order Crocodilia — Alligators and crocodiles (Triassic to Recent)
      - Order Pterosauria — Flying and gliding reptiles (Triassic to Cretaceous)
      - *Superorder Dinosauria*
        - Order Saurischia — "Lizard-hipped" dinosaurs (Triassic to Cretaceous)
        - Order Ornithischia — "Bird-hipped" dinosaurs (Triassic to Cretaceous)
  - *Subclass Euryapsida* — Aquatic diapsids which lost the lower temporal opening, including ichthyosaurs, nothosaurs, placodonts, and plesiosaurs (Triassic to Cretaceous)
  - *Subclass Synapsida* — Mammal-like reptiles with a single temporal opening
    - Order Pelycosauria — Early synapsids, including "sail-backed" reptiles (Pennsylvanian to Permian)
    - Order Therapsida — Later mammal-like reptiles (Permian to Triassic)
- Class Aves — Birds (Jurassic to Recent)
  - *Subclass Archaeornithes* — Primitive toothed birds (Jurassic to Cretaceous)
  - *Subclass Ornithurae* — Advanced Cretaceous and Cenozoic birds
- Class Mammalia
  - *Subclass Eotheria* — Very primitive mammals including morganucodonts (Triassic to Jurassic)
  - *Subclass Allotheria* — Early mammals with complex teeth, including multituberculates (Jurassic to Eocene)

Subclass Prototheria	Egg-laying mammals, including monotremes (Triassic to Recent)
Subclass Theria	Most mammals
Infraclass Pantotheria	Group ancestral to most modern mammals; includes eupantotheres (Jurassic to Cretaceous)
Infraclass Metatheria	Pouched mammals, or marsupials (Cretaceous to Recent)
Infraclass Eutheria	Placental mammals (Cretaceous to Recent)
Order Insectivora	Insect-eating mammals such as shrews, moles, and hedgehogs (Cretaceous to Recent)
Order Edentata	Glyptodonts and ground sloths and modern armadillos, anteaters and tree sloths (Paleocene to Recent)
Order Rodentia	Herbivores with enlarged incisor teeth, including beavers, mice, and squirrels (Paleocene to Recent)
Order Lagomorpha	Gnawing mammals including rabbits, hares, and pikas (Paleocene to Recent)
Order Carnivora	Most modern carnivorous mammals, including cats, dogs, sea lions, and bears (Paleocene to Recent)
Order Cetacea	Whales and porpoises (Eocene to Recent)
Order Artiodactyla	Even-toed hoofed mammals, including antelope, camels, deer, giraffes, hippopotamuses, and pigs (Eocene to Recent)
Order Perissodactyla	Odd-toed hoofed mammals, including chalicotheres and titanotheres and modern horses, rhinoceroses, and tapirs (Eocene to Recent)
Order Proboscidea	Mammoths, mastodons, and modern elephants (Eocene to Recent)
Order Primates	Apes, lemurs, monkeys, tarsiers, and humans (Paleocene to Recent)

Appendix G

Sources of Geologic Literature, Photographs, and Maps

American Geological Institute (AGI)
4220 King Street
Alexandria, VA 22302-1502
http://www.agiweb.org
or
AGI Publications Center
P.O. Box 205
Annapolis Junction, MD 20701
http://www.agiweb.org/pubserv.html

Issues the monthly "Bibliography and Index of Geology," which includes worldwide references and contains listings by author and subject.

Canada Centre for Remote Sensing
588 Booth Street
Ottawa, Ontario K1A-0Y7
CANADA
http://www.ccrs.nrcan.gc.ca/

Source for obtaining aerial photographs and satellite imagery of Canada.

Earth Science Information Center
U.S. Geological Survey
507 National Center
Reston, VA 20192
http://mapping.usgs.gov/esic/esic.html

Publishes topographic and other scale maps of the United States. Indexes listing available map scales are available for all the states.

Earthquake Engineering Research Institute
499 14th Street, Suite 320
Oakland, CA 94612-1934
http://www.eeri.org

Provides a valuable field guide for learning about and studying earthquakes.

Federal Emergency Management Agency
P.O. Box 2012
Jessup, MD 20794-2012
ATTN: Publications
http://www.fema.gov

Provides a number of publications aimed at helping citizens prepare for geologic disasters, including "Are you Ready? Your Guide to Disaster Preparedness Checklist." Also issues a general "FEMA Publications Catalog."

Geological Society of America
3300 Penrose Place
P.O. Box 9140
Boulder, CO 80301-9140
http://www.geosociety.org

Publishes a large and diverse number of papers and journals concerning the geology and mineral resources of North America.

National Air Photo Library
615 Booth Street
Room 180
Ottawa, Ontario K1A-0E9
CANADA
http://napl.ccm.nrcan.gc.ca/collection.html

Source for obtaining aerial photographs and satellite imagery of Canada.

National Climatic Data Center (NCDC)
Federal Building
151 Patton Avenue
Asheville, NC 28801-5001
http://www.ncdc.noaa.gov

Issues a monthly publication of climatological data, such as temperature and precipitation statistics for any given state or region. The NCDC maintains up-to-date weather records for the entire United States.

National Geophysical and Solar–Terrestrial Data Center (part of the National Oceanic and Atmospheric Administration)
325 Broadway
Boulder, CO 80303
http://www.noaa.gov

Maintains worldwide computer files of earthquakes recorded by seismographs and historic earthquake data.

National Oceanic and Atmospheric Administration (NOAA)
Outreach Unit
1305 East West Highway
Rm. 1W204
Silver Spring, MD 20910
http://www.noaa.gov

This federal agency maintains a large collection of aerial photographs of the coastal regions of the United States. A detailed index is available.

State geological surveys

All states maintain a geological survey that produces technical reports, maps, and other publications at the state or county level.

U.S. Department of Agriculture
Aerial Photography Field Office
2222 West 2300 South
Salt Lake City, UT 84119-2020
http://www.fsa.usda.gov/dam/APFO/airfto.htm

Can provide aerial photographs of most of the United States. A variety of scales is available.

U.S. Geological Survey (USGS)
Information Services
Box 25286
Building 810
Denver Federal Center
Denver, CO 80225
http://www.usgs.gov

Produces a large number of maps, reports, circulars, professional papers, bulletins, water resources publications, and so on. Circular number 777 of the USGS provides an annually updated guide to obtaining information on the earth sciences.

U.S. Geological Survey (USGS)
EROS Data Center
47914 252nd Street
Sioux Falls, SD 57198-0001
http://edcwww.cr.usgs.gov

Provides computer listings of most satellite imagery, high-altitude aerial photography, and photographs obtained during the Apollo, Skylab, and Gemini space missions. Worldwide coverage and indexes are available.

Appendix H

Careers in the Geosciences

Careers in the geosciences are numerous and varied. Several have been alluded to throughout this book. Employers in these fields include the energy industry, which hires about half of all professional geoscientists; the mining industry; federal and state governments; consulting firms, especially in environmental issues and hydrogeology; and academia, which employs more than 10% of all geoscientists.

Each of these broad fields, of course, involves a number of subdisciplines: the energy industry, for example, requires expertise in sedimentology and stratigraphy; mining careers require a background in economic geology, petrology, mineralogy, crystallography, or structural geology. The following is a list of some of the occupations in which people with an interest in the geosciences are employed:

Economic geologists conduct field investigations to determine the locations and economic viability of mineral deposits; they also investigate the genesis of mineral deposits. Mining companies typically employ economic geologists.

Engineering geologists investigate the geologic factors that affect human-made structures such as bridges, dams, and buildings, and the geologic effects of mass wasting.

Environmental geologists work in assessing, solving, and preventing problems associated with the pollution of soil, bedrock, and groundwater. For example, they assist in the selection of suitable locations for municipal- and hazardous-waste facilities such as landfills and waste-storage facilities, and may also help design such facilities.

Geochemists investigate the nature and distribution of chemical elements in the geologic environment.

Geochronologists determine the age of geologic materials, helping to reconstruct the geologic history of the Earth.

Geomorphologists study landforms, the rates and intensity of processes that created them, and the relationship of these landforms to the underlying geologic structures and climates that existed during their evolution.

Geophysicists attempt to determine the internal structure and properties of the Earth. They may focus on specific factors such as seismic waves, geomagnetism, or gravity.

Glaciologists investigate the physical and chemical properties of glacial masses. They also study the development, movement, and decay of glaciers and ice sheets and their deposits.

Hydrogeologists study the production, distribution, movement, and quality of groundwater in the Earth's crust. Many hydrogeologists work in the environmental industry, in cooperation with environmental geologists.

Hydrologists study the distribution, movement, and quality of surface bodies of water.

Marine geologists investigate the topography and sediments of the world's oceans. Marine geologists may work closely with petroleum geologists in off-shore exploration projects. They often work closely with oceanographers.

Mineralogists study the formation, composition, and properties—both physical and chemical—of minerals.

Paleontologists collect fossils and determine their age, reconstruct past environments, and regionally correlate rocks by determining the evolutionary sequence of fossil assemblages found in the rocks. Many paleontologists work in the oil industry.

Petroleum geologists work in the oil industry, and are involved with the exploration and production of petroleum and natural gas. They may also research unconventional sources such as oil shales and sands.

Petrologists study the mineralogical relationships of rocks, and specialize in determining the genesis of rocks; they may focus on either sedimentary, igneous, or metamorphic petrology.

Planetary geologists study the planets and their satellites to better understand the evolution of the solar system. Most planetary geologists are employed by universities or advanced research organizations such as NASA (National Aeronautics and Space Administration).

Sedimentologists study the formation of sedimentary rocks by assessing the processes of transportation, erosion, and deposition. Many sedimentologists work in the petroleum industry.

Stratigraphers decipher the sequence of rocks using the principle of superposition. They study the time and space relationships of rock sequences in an effort to determine the geologic history of an area.

Structural geologists investigate the phenomena and structures produced by the deformation of the Earth's crust, such as faults and folds. Most of the data used in structural geology are collected during detailed field work.

An academic background that includes courses in math and other sciences as well as in the specific geological fields mentioned here would be practical preparation for a geoscience career.

Glossary

ablation zone See *zone of ablation.*

abrasion A form of *mechanical weathering* that occurs when loose fragments or particles of rocks and minerals that are being transported, as by water or air, collide with each other or scrape the surfaces of stationary rocks.

Absaroka sequence The last of the major *cratonic sequences* (see glossary definition) deposited during the *Paleozoic Era.* This sequence consists of *Pennsylvanian* and *Permian* age sediments deposited by *transgression* and *regression* on the platform and is bounded above and below by *unconformities* in many places.

absolute dating The fixing of a geological structure or event in time, as by counting tree rings. See also *relative dating.*

Acadian orogeny Mountain-building event in the northern Appalachians occurring during the *Devonian Period* resulting from the collision of the Avalonian terrane with the North American continent. The Catskill Mountains of New York are the *erosional* debris from the mountains formed by this collision.

accumulation zone See *zone of accumulation.*

acid rain Rain that contains such acidic compounds as sulfuric acid and nitric acid, which are produced by the combination of atmospheric water with oxides released when *hydrocarbons* are burned. Acid rain is widely considered responsible for damaging forests, crops, and human-made structures, and for killing aquatic life.

acritarch An early eukayote fossil, first appearing during the Middle *Proterozoic,* that resembles the tough, outer coating grown by some types of unicellular planktonic algae. Some acritarchs may be the remains of dinoflagellates.

adaptive radiation The process of forming diverse *species* from a common ancestor in response to different ecological pressures.

aeration zone See *zone of aeration.*

Allegheny orogeny The final stage in the building of the Appalachians resulting from the late *Paleozoic* collision of Africa and North America. This event is most responsible for folding and faulting in the central and southern Appalachians.

alluvial fan A triangular deposit of *sediment* left by a stream that has lost velocity upon entering a broad, relatively flat valley.

alluvium A deposit of *sediment* left by a stream in the stream's channel or on its *floodplain.*

alpine glacier A mountain *glacier* that is confined by highlands.

Alpine orogen The mountainous region of southern Europe, stretching from Spain to Turkey, formed by closing of the *Tethys Sea.*

ammonites An extinct group of *cephalopods* with coiled conchs characterized by folded septa between the chambers.

amniote egg An egg in which the amniotic fluid is enclosed by a tough membrane or shell, characteristic of reptiles and birds.

Ancestral Rockies A collection of mountain ranges formed by uplift of blocks of *crust* in the southwestern United States during the *Pennsylvanian Period.* The locations of some of these ancient ranges corresponds to parts of the modern Rockies.

Andean Cordillera The mountainous western side of South America formed by the *Cenozoic* subduction of the *Farallon* plate, and later the Nazca plate.

andesite The dark, aphanitic, *extrusive rock* that has a *silica* content of about 60% and is the second most abundant volcanic rock. Andesites are found in large quantities in the Andes Mountains.

andesite line The geographic boundary between the *basalts* and *gabbros* of the Pacific Ocean basin and the *andesites* at the subductive margins of the surrounding continents.

angiosperms Plants whose reproductive structure includes flowers, the "flowering plants," first appearing during the *Mesozoic.*

angle of repose The maximum angle at which a pile of unconsolidated material can remain stable.

ankylosaurs A group of quadrupedal *ornithischians* that were low-built and had backs covered with bony plates.

anthracite A hard, jet-black coal that develops from *lignite* and *bituminous coal* through *metamorphism,* has a carbon content of 92% to 98%, and contains little or no gas. Anthracite burns with an extremely hot, blue flame and very little smoke, but it is difficult to ignite and both difficult and dangerous to mine.

anthropoids Advanced primates with more upright posture and flattened faces, consists of the Old World monkeys, New World monkeys, and the hominoids.

anticline A convex *fold* in rock, the central part of which contains the oldest section of rock. See also *syncline.*

Antler orogeny A *Devonian* to *Mississippian* mountain-building event in the West caused by collision of an island arc and North America.

aquiclude An impermeable body of rock that may absorb water slowly but does not transmit it.

aquifer A *permeable* body of rock or *regolith* that both stores and transports groundwater.

archaeocyathids Vase-shaped animals with calcareous skeletons, possibly related to sponges, that built reefs during the Early *Cambrian.*

Archaeopteryx The oldest generally accepted fossil bird, which still possessed many reptilian characteristics, from the Upper *Jurassic.*

Archean Eon The first eon of the Precambrian for which there is a clear geologic record, lasting from 3.8 billion to 2.5 billion years ago.

arête A sharp ridge of erosion-resistant rock formed between adjacent *cirque glaciers.*

aridity index The ratio of a region's potential annual evaporation, as determined by its receipt of solar radiation, to its average annual precipitation.

arroyo A small, deep, usually dry channel eroded by a short-lived or intermittent desert stream.

artesian Of, being, or concerning an *aquifer,* in which water rises to the surface due to pressure from overlying water.

asthenosphere A layer of heat-softened but solid, mobile rock comprising the lower part of the upper *mantle* from about 100 to 350 kilometers beneath the Earth's surface. See also *lithosphere.*

Atlantic Margin The passive eastern edge of North America, formed by rifting from Africa during the *Mesozoic Era.*

atoll A circular reef that encloses a relatively shallow lagoon and extends from a very great depth to the sea surface. An atoll forms when an oceanic island ringed by a *barrier reef* sinks below sea level.

atom The smallest particle that retains all the chemical properties of a given *element.*

atomic mass 1. The sum of *protons* and *neutrons* in an atom's nucleus. 2. The combined mass of all particles in a given atom.

atomic number The number of *protons* in the nucleus of a given atom. *Elements* are distinguished from each other by their atomic numbers.

aureole A section of rock that surrounds an intrusion and shows the effects of *contact metamorphism.*

australopithecines Early *hominids* (see glossary definition) that lived from about 4.2 million years ago to about 1.4 million years ago, much smaller than modern humans.

backswamp A wetland area formed when flood water fills surface depressions in a stream's *floodplain.*

Baltica A late *Proterozoic* to early *Paleozoic* continent consisting of most of northern Europe, formed by the breakup of *Rodinia* (see glossary definition).

banded iron formation A rock consisting of alternating layers of dark iron-oxide minerals and red chert, formed during the *Archean* and Early *Proterozoic* by the combination of oxygen with iron dissolved in the ocean.

barchan dune A crescent-shaped *dune* that forms around a small patch of vegetation, lies perpendicular to the prevailing wind direction, and has a gentle, convex windward slope and a steep, concave leeward slope. Barchan dunes typically form in arid, inland deserts with stable wind direction and relatively little sand.

barrier island A ridge of sand that runs parallel to the main coast but is separated from it by a bay or lagoon. Barrier islands range from 10 to 100 kilometers in length and from 2 to 5 kilometers in width. A barrier island may be as high as 6 meters above sea level.

barrier reef A long, narrow reef that runs parallel to the main coast but is separated from it by a wide lagoon.

basal sliding The process by which a *glacier* undergoes thawing at its base, producing a film of water along which the glacier then flows. Basal sliding primarily affects glaciers in warm climates or mid-latitude mountain ranges.

basalt The dark, dense, aphanitic, *extrusive rock* that has a silica content of 40% to 50% and makes up most of the ocean floor. Basalt is the most abundant volcanic rock in the Earth's crust.

base level The lowest level to which a *stream* can erode the channel through which it flows, generally equal to the prevailing global sea level.

basin A round or oval depression in the Earth's surface, containing the youngest section of rock in its lowest, central part. See also *dome.*

Basin and Range province A region of the *Cordillera* characterized by elongate, narrow mountain ranges separated by broad valleys, formed by late *Cenozoic* extension of the crust.

batholith A massive discordant *pluton* with a surface area greater than 100 square kilometers, typically having a depth of about 30 kilometers. Batholiths are generally found in elongated mountain ranges after the country rock above them has eroded.

baymouth bar A narrow ridge of sand (a *spit*) that stretches completely across the mouth of a bay. (Also called *bay bar* and *bay barrier.*)

beach The part of a *coast* that is washed by waves or tides, which cover it with *sediments* of various sizes and composition, such as sand or pebbles.

beach drift 1.The process by which a *longshore current* moves *sediments* along a beach face. 2. The sediments so moved. Beach drift typically consists of sand, gravel, shell fragments, and pebbles. See also *longshore drift.*

bedding The division of *sediment* or *sedimentary rock* into parallel layers (beds) that can be distinguished from each other by such features as chemical composition and grain size.

bed load A body of coarse particles that tend to move along the bottom of a *stream.*

belemnites Squid-like *cephalopods* with a straight internal shell that lived during the *Mesozoic* and earliest *Cenozoic.*

biogenic chert A type of *sedimentary rock* composed of silica-based organic debris such as the shells and skeletons of small marine organisms. Unlike *inorganic chert,* which forms in the shape of nodules, organic chert typically forms in layers.

biogenic limestone A type of *sedimentary rock* composed mostly of calcite (calcium carbonate), formed from the shells and skeletons of small marine organisms. See also *inorganic limestone.*

biomass fuel A renewable fuel derived from a living organism or the by-product of a living organism. Biomass fuels include wood, dung, methane gas, and grain alcohol.

bitumen Any of a group of solid and semi-solid *hydrocarbons* that can be converted into liquid form by heating. Bitumens can be refined to produce such commercial products as gasoline, fuel oil, and asphalt.

bituminous coal A shiny black coal that develops from deeply buried *lignite* through heat and pressure, and that has a carbon content of 80% to 93%, which makes it a more efficient heating fuel than lignite.

bivalves The *class* of *mollusks* that lives within a two-valved shell; includes clams and oysters.

blowout A small area of land that has been somewhat lowered due to *deflation,* usually where stabilizing surface vegetation has been previously disturbed.

body wave A type of *seismic wave* that transmits energy from an earthquake's *focus* through the Earth's interior in all directions. See also *surface wave.*

bond To combine, by means of chemical reaction, with another atom to form a *compound.* When an atom bonds with another, it either loses, gains, or shares electrons with the other atom.

Bowen's reaction series The sequence of *igneous rocks* formed from a mafic *magma,* assuming mineral crystals that have already formed continue to react with the liquid magma and so evolve into new minerals, thereby creating the next rock in the sequence.

brachiopods A *phylum* of marine invertebrates possessing the lophophore structure that live within a two-valved shell. They were the most common inhabitants of the *Paleozoic* sea floor.

braided stream A network of converging and diverging *streams* separated from each other by narrow strips of sand and gravel.

breakwater A wall built seaward of a coast to intercept incoming waves and so protect a harbor or shore. Breakwaters are typically built parallel to the coast.

breccia A *clastic rock* composed of particles more than 2 millimeters in diameter and marked by the angularity of its component grains and rock fragments.

brittle failure The rupturing of rock, a type of permanent deformation caused by great stress under relatively low temperature and pressure conditions. See also *plastic deformation.*

bryozoans A *phylum* of small colonial animals possessing the lophophore structure.

Burgess Shale fauna A diverse assemblage of hard- and soft-bodied animals preserved in formations of Middle *Cambrian*-age, including the Burgess Shale.

burial metamorphism A form of *regional metamorphism* that acts on rocks covered by 5 to 10 kilometers of rock or sediment, caused by heat from the Earth's interior and *lithostatic pressure.*

caldera A vast depression at the top of a *volcanic cone,* formed when an eruption substantially empties the reservoir of *magma* beneath the cone's summit. Eventually the summit collapses inward, creating a caldera. A caldera may be more than 15 kilometers in diameter and more than 1000 meters deep.

Caledonian orogeny The collision and suturing of *Baltica* and *Laurentia* during the *Silurian* and *Devonian* periods, forming a mountain belt that extended from Canada to Greenland, northern Great Britain, and Scandinavia.

Cambrian explosion The rapid appearance of most modern animal phyla during the Early *Cambrian Period.*

Cambrian Period The earliest period of the *Paleozoic Era,* from 544 to 505 million years ago, defined by the stratigraphically lowest occurrence of complex trace fossils.

capacity The ability of a given *stream* to carry *sediment,* measured as the maximum quantity it can transport past a given point on the channel bank in a given amount of time. See also *competence.*

carbon-14 dating A form of *radiometric dating* that relies on the 5730-year half-life of radioactive carbon-14, which decays into nitrogen-14, to determine the age of rocks in which carbon-14 is present. Carbon-14 dating is used for rocks from 100 to 100,000 years old.

carbonate One of several minerals containing one central carbon atom with strong *covalent bonds* to three oxygen atoms and typically having *ionic bonds* to one or more positive ions.

Carboniferous Period A late *Paleozoic* period beginning 360 million years ago and ending 286 million years ago. In the United States, the *Mississippian* and *Pennsylvanian* periods are used instead of the Lower and Upper Carboniferous used elsewhere.

carnivores Flesh-eating placental mammals, appearing during the Paleocene Epoch; includes cats, dogs, otters, and bears.

Cascade Range An active volcanic arc formed over the last 2 million years by subduction of the *Juan de Fuca plate.*

catastrophism The hypothesis that a series of immense, brief, worldwide upheavals changed the Earth's curst greatly and can account for the development of mountains, valleys, and other features of the Earth. See also *uniformitarianism.*

cave An opening beneath the surface of the Earth, a *karst* feature generally produced by *dissolution* of carbonate bedrock.

cementation The process by which loose sediment grains are bound together by *precipitated* minerals originally dissolved during the *chemical weathering* of preexisting rocks.

Cenozoic Era The latest era of the *Phanerozoic Eon,* following the *Mesozoic Era* and continuing to the present time, and marked by the presence of a wide variety of mammals, including the first hominids. It is divided into two periods, the Tertiary and the Quaternary.

cephalopods A *class* of *mollusks* characterized by a prominent head and division of the foot into multiple arms or tentacles. Includes the modern squid, octopus, and nautilus as well as the ancient ammonoids.

ceratopsians A group of quadrupedal *ornithischians* characterized by a bony frill covering the neck and facial horns.

chemical sediment *Sediment* that is composed of previously dissolved minerals that have either precipitated from water or been extracted from water by living organisms and deposited when the organisms died or discarded their shells.

chemical weathering The process by which chemical reactions alter the chemical composition of rocks and minerals that are unstable at the Earth's surface and convert them into more stable substances; *weathering* that changes the chemical makeup of a rock or mineral. See also *mechanical weathering.*

Chicxulub structure A buried circular depression, up to 180 kilometers wide, located on the northern coast of the Yucatan Peninsula of Mexico, generally believed to be the crater of an asteroid impact at the *Cretaceous-Tertiary* boundary.

Chinle Group A *stratigraphic* unit most famous for deposits of petrified wood, consisting of *alluvial sediments* deposited over a broad region of the American West during the Late *Triassic.*

chromosomes Elongate strands containing the cellular *DNA* (see glossary definition) located in the nucleus of eukaryotes.

cinder cone A *pyroclastic cone* composed primarily of cinders.

cirque A deep, semi-circular basin eroded out of a mountain by an *alpine glacier.*

cirque glacier A small *alpine glacier* that forms inside a *cirque,* typically near the head of a valley.

class The primary division of *phylum* in the Linnaean system of classification. A class consists of one or more *orders.*

clastic Being or pertaining to a *sedimentary rock* composed primarily from fragments of preexisting rocks or fossils.

coal A member of a group of easily combustible, organic *sedimentary rocks* composed mostly of plant remains and containing a high proportion of carbon.

coast The area of dry land that borders on a body of water.

coccolithophorids A golden-brown variety of algae whose calcareous shells are an important component of *Cretaceous* chalks.

col A mountain pass eroded out of a ridge by an *alpine glacier.*

Colorado Plateau A block of relatively undeformed crust encompassing northern Arizona, eastern Utah, western Colorado, and northwestern New Mexico that was uplifted during the late *Cenozoic.*

Columbia Plateau A region of eastern Washington and Oregon and western Idaho that was covered by flood basalts during the *Miocene Epoch.*

compaction The process by which the volume or thickness of sediment is reduced due to pressure from overlying layers of sediment.

competence The ability of a given *stream* to carry *sediment,* measured as the diameter of the largest particle that the stream can transport. See also *capacity.*

composite cone See *stratovolcano.*

compound Two or more *elements* bonded together in specific, constant proportions. A compound typically has physical characteristics different from those of its constituent elements.

compression *Stress* that reduces the volume or length of a rock, as that produced by the *convergence* of plate margins.

cone of depression An area in a *water table* along which water has descended into a well to replace water drawn out, leaving a gap shaped like an inverted cone.

confining pressure See *lithostatic pressure.*

conglomerate A *clastic* rock composed of particles more than 2 millimeters in diameter and marked by the roundness of its component grains and rock fragments.

contact metamorphism *Metamorphism* that is caused by heat from a magmatic intrusion.

continental collision The *convergence* of two continental plates, resulting in the formation of mountain ranges.

continental drift The hypothesis, proposed by Alfred Wegener, that today's continents broke off from a single supercontinent and then plowed through the ocean floors into their present positions. This explanation of the shapes and locations of Earth's current continents evolved into the theory of *plate tectonics.*

continental ice sheet An unconfined *glacier* that covers much or all of a continent.

continental platform That portion of a continent where the *continental shield* is covered with a veneer of younger rock. The continental shield and the continental platform together constitute a continent's tectonically stable *craton.*

continental shield A broad area of exposed crystalline rock in a continental interior that is the oldest part of the continent (e.g., the Canadian Shield in North America). The continental shield together with the *continental platform* constitute a continent's tectonically stable *craton.*

convection cell The cyclical movement of material in the *asthenosphere* that causes the plates of the *lithosphere* to move. Heated material becomes less dense and rises toward the solid lithosphere, through which it cannot rise further and therefore begins to move horizontally, dragging the lithosphere along with it and pushing forward the cooler, denser material in its path. The cooler material eventually sinks down lower into the mantle, becoming heated there and rising up again, continuing the cycle. See also *plate tectonics.*

convergence The coming together of two lithospheric plates. Convergence causes *subduction* when one or both plates is oceanic, and mountain formation when both plates are continental. See also *divergence.*

Cordillera The mountainous region of western North America.

Cordilleran ice sheet A continental ice sheet that covered the Rockies and Coast Ranges in western Canada during the *Pleistocene Epoch.*

Cordilleran orogeny A prolonged sequence of mountain-building events in western North America during the *Mesozoic* to early *Cenozoic* consisting of the *Nevadan orogeny,* the *Sevier orogeny,* and the *Laramide orogeny* during the Late *Cretaceous* to Early *Tertiary* periods.

core The innermost layer of the Earth, consisting primarily of pure metals such as iron and nickel. The core is the densest layer of the Earth, and is divided into the outer core, which is believed to be liquid, and the inner core, which is believed to be solid. See also *crust* and *mantle.*

correlation The process of determining that two or more geographically distant rocks or rock strata originated in the same time period.

covalent bond The combination of two or more atoms by sharing electrons so as to achieve chemical stability. Atoms that form covalent bonds generally have outer energy levels containing three, four, or five electrons. Covalent bonds are generally stronger than other bonds.

crater See *volcanic crater.*

craton That region of a continent, consisting of the *continental shield* and the *continental platform,* that has been tectonically stable for a vast period of time and contains its oldest rocks.

cratonic sequences Large-scale packages of sedimentary formations, separated by *unconformities,* formed by major *transgressions* and *regressions* on the North American *craton.*

creep The slowest form of *mass movement,* measured in millimeters or centimeters per year. It affects unconsolidated materials such as soil or *regolith,* the particles of which are continuously rearranged by gravity on virtually any slope.

Cretaceous Period The final period of the *Mesozoic Era,* lasting from 140 million years ago until 66 million years ago.

crinoids A group of echinoderms with stalked, segmented bodies. Ancient crinoids lived anchored to the ocean floor.

Cro-Magnon Nomadic European group that existed from about 35,000 years ago to about 10,000 years ago, characterized by the manufacture of tools and decorative arts.

cross-bed A bed made up of particles dropped from a moving current, as of wind or water, and marked by a downward slope that indicates the direction of the current that deposited them.

cross-cutting relationships See *principle of cross-cutting relationships.*

crust The outermost layer of the Earth, consisting of relatively low-density rocks. See also *core* and *mantle.*

crystal A mineral in which the systematic internal arrangement of *atoms* is outwardly reflected as a latticework of repeated three-dimensional units that form a geometric solid with a surface consisting of symmetrical planes.

crystal structure 1. The geometric pattern created by the systematic internal arrangement of atoms in a mineral. 2. The systematic internal arrangement of atoms in a mineral. See also *crystal.*

cyanobacteria A photosynthetic type of bacteria whose fossils have been found in rocks dated to 3.5 billion years, the oldest known fossils. Mats of these bacteria can build layered structures called *stromatolites* (see glossary definition).

cycads A group of *gymnosperms* with cylindrical trunks and large fern-like leaves, most common during the *Mesozoic,* but still surviving today.

cyclothems Orderly, repetitive sequences of beds consisting of alternating marine and nonmarine sediments, deposited on the *craton* during the *Pennsylvanian Period.*

daughter isotope An *isotope* that forms from the radioactive decay of a *parent isotope.* A daughter isotope may or may not be of the same element as its parent. If the daughter isotope is radioactive, it will eventually become the parent isotope of a new daughter isotope. The last daughter isotope to form from this process will be stable and nonradioactive.

debris avalanche The sudden, extremely rapid *mass movement* downward of entire layers of *regolith* along very steep slopes. Debris avalanches are generally caused by heavy rains.

debris flow 1. The rapid, downward *mass movement* of particles coarser than sand, often including boulders 1 meter or more in diameter, at a rate ranging from 2 to 40 kilometers per hour. Debris flows occur along fairly steep slopes. 2. The material that descends in such a flow.

deflation The process by which wind erodes a surface by picking up and transporting loose rock particles.

deflation basin A large-scale *blowout,* occurring where local bedrock is particularly soft or has been extensively crushed by faulting.

delta An *alluvial fan* having its apex at the mouth of a *stream.*

desert A region with an average annual rainfall of 10 inches or less and sparse vegetation, typically having thin, dry, and crumbly soil. A desert has an *aridity index* greater than 4.0.

desertification The process through which a desert takes over a formerly nondesert area. When a region begins to undergo desertification, the new conditions typically include a significantly lowered *water table,* a reduced supply of surface water, increased salinity in natural waters and soils, progressive destruction of native vegetation, and an accelerated rate of erosion.

desert pavement A closely packed layer of rock fragments concentrated in a layer along the Earth's surface by the *deflation* of finer particles.

detrital sediment *Sediment* that is composed of transported solid fragments of preexisting igneous, sedimentary, or metamorphic rocks.

Devonian Period A period of the middle *Paleozoic Era* lasting from 408 million years ago until 360 million years ago.

diapsids A group of reptiles that appeared during the late Paleozoic, distinguished by two openings behind the eye socket.

dike A discordant tabular *pluton* that is substantially wider than it is thick. Dikes are often steeply inclined or nearly vertical. See also *sill.*

dilatancy The expansion of a rock's volume caused by *stress* and deformation.

diorite Any of a group of dark, phaneritic, *intrusive rocks* that are the *plutonic* equivalents of *andesite.*

dip The angle formed by the inclined plane of a geological structure and the horizontal plane of the Earth's surface.

dip-slip fault A *fault* in which two sections of rock have moved apart vertically, parallel to the *dip* of the fault plane.

directed pressure Force exerted on a rock along one plane, flattening the rock in that plane and lengthening it in the perpendicular plane.

disappearing stream A surface *stream* that drains rapidly and completely into a *sinkhole.*

discharge The volume of a *stream's* water that passes a given point per unit of time.

displaced terrane A body of foreign rock that has been carried from elsewhere by plate motion and attached to a continent's coast by collision.

dissolution A form of *chemical weathering* in which water molecules, sometimes in combination with acid or another compound in the environment, attract and remove oppositely charged ions or ion groups from a mineral or rock.

dissolved load A body of sediment carried by a *stream* in the form of *ions* that have dissolved in the water.

distributary One of a network of small *streams* carrying water and sediment from a *trunk stream* into an ocean.

divergence The process by which two lithospheric plates separated by *rifting* move farther apart, with soft mantle rock rising between them and forming new oceanic *lithosphere.* See also *convergence.*

DNA Deoxyribonucleic acid, a molecule consisting of alternating sugar and phosphate groups and nitrogen-based nucleotides, containing the genetic information of an organism.

dolostone A *sedimentary rock* composed primarily of dolomite, a mineral made up of calcium, magnesium, carbon, and oxygen. Dolostone is thought to form when magnesium ions replace some of the calcium ions in limestone, to which dolostone is similar in both appearance and chemical structure.

dome A round or oval bulge on the Earth's surface, containing the oldest section of rock in its raised, central part. See also *basin.*

drainage basin The area from which water flows into a *stream.* Also called a *watershed.*

drainage divide An area of raised, dry land separating two adjacent *drainage basins.*

drumlin A long, spoon-shaped hill that develops when pressure from an overriding *glacier* reshapes a *moraine.* Drumlins range in height from 5 to 50 meters and in length from 400 to 2000 meters. They slope down in the direction of the ice flow.

dryomorphs Early ape-like creatures that appeared in Africa during the Early *Miocene,* distinguished by a long, monkeylike torso and limbs, and an ape-like skull, jaws, and teeth.

dune A usually asymmetrical mound or ridge of sand that has been transported and deposited by water or wind. Dunes form in both arid and humid climates.

dynamothermal metamorphism A form of *regional metamorphism* that acts on rocks caught between two *converging* plates and is initially caused by *directed pressure* from the plates, which causes some of the rocks to rise and others to sink, sometimes by tens of kilometers. The rocks that fall then experience further dynamothermal metamorphism, this time caused by heat from the Earth's interior and *lithostatic pressure* from overlying rocks.

earthflow 1. The *flow* of a dry, highly viscous mass of claylike or silty *regolith,* typically moving at a rate of 1 or 2 meters per hour. 2. The material that descends in such a flow.

earthquake A movement within the Earth's *crust* or *mantle,* caused by the sudden rupture or repositioning of underground rocks as they release *stress.*

East African rift Part of a three-way rift, a chain of rift valleys formed during the *Pliocene* by stretching the crust.

echinoderms A phylum consisting of five classes of marine creatures: sea urchins, starfish, brittle stars, crinoids, and sea cucumbers, covered by calcareous plates or spines.

ectotherms Animals with no internal ability to regulate body temperature.

edentates Placental mammals with few or no teeth, first appearing during the *Paleocene,* including modern armadillos, tree sloths, and anteaters.

Ediacaran fauna The Late *Proterozoic* impressions of soft-bodied organisms representing the oldest-known large metazoans. Many may be the ancestors of modern animal groups (sponges, mollusks, anemones), whereas others are of uncertain classification.

elastic deformation A temporary change in the shape or volume of a rock, caused by less stress than that which produces *brittle failure* or *plastic deformation.*

electron A negatively charged particle that orbits rapidly around the *nucleus* of an *atom.* See also *proton.*

element A form of matter that cannot be broken down into a chemically simpler form by heating, cooling, or chemical reactions. There are 106 known elements, 92 of them natural and 14 synthetic. Elements are represented by one- or two-letter abbreviations. See also *atom, atomic number.*

endosymbiosis Theory for the origin of eukaryotes by which larger prokaryotes enfolded within their cell membranes smaller prokaryotes that performed useful functions and formed symbiotic relationships.

endotherms Animals with a metabolism that allows for internal control of body temperature.

energy level The path of a given electron's orbit around a nucleus, marked by a constant distance from the nucleus.

Eocene Epoch An epoch of the early *Tertiary Period,* lasting from 58 million years ago to 38 million years ago.

eons The largest time spans shown on the geologic time scale, designating the major developments (mostly biological) in Earth history. See also *Hadean Eon, Archean Eon, Proterozoic Eon,* and *Phanerozoic Eon.*

epeiric seas Shallow seaways covering large portions of the *cratons.*

epicenter The point on the Earth's surface that is located directly above the *focus* of an *earthquake.*

epochs The time spans that subdivide *periods* in the *geologic time scale.*

eras The second-largest time spans shown on the *geologic time scale* (subdividing the *eons*), defined by their dominant life forms. See also *Paleozoic Era, Mesozoic Era,* and *Cenozoic Era.*

erosion The process by which particles of rock and soil are loosened, as by *weathering,* and then transported elsewhere, as by wind, water, ice, or gravity.

esker A ridge of *sediment* that forms under a glacier's *zone of ablation,* made up of sand and gravel deposited by meltwater. An esker may be less than 100 meters or more than 500 kilometers long, and may be anywhere from 3 to over 300 meters high.

eukaryotic Having a cell *nucleus.* Describes the more complex life forms that evolved from the Earth's first simple *prokaryotic* organisms as well as all higher organisms that evolved subsequently.

eupantotheres A group of *Jurassic* mammals that are thought to be ancestral to the marsupial and placental mammals.

eustatic Changes in sea level that can be demonstrated as occurring globally resulting from major tectonic cycles, sea-floor spreading, or major episodes of *glaciation.*

evaporite An inorganic *chemical sediment* that *precipitates* when the salty water in which it had dissolved evaporates.

even-toed ungulates Diverse group of hoofed mammals that includes deer, giraffes, bison, pigs, and hippopotamuses.

extrusive rock An *igneous rock* formed from *lava* that has flowed out onto the Earth's surface, characterized by rapid solidification and grains that are so small as to be barely visible to the naked eye.

fall The fastest form of *mass movement,* occurring when rock or sediment breaks off from a steep or vertical slope and descends at a rate of 9.8 meters per second. A fall can be extremely dangerous.

family The primary division of orders in the Linnaean system of biological classification, consisting of one or more genera (plural of *genus*).

Farallon plate The ocean plate bordering North and South America during the *Mesozoic* and most of the *Cenozoic.* Most of this plate has been *subducted,* leaving the smaller Juan de Fuca and Cocos plates.

fault A fracture dividing a rock into two sections that have visibly moved relative to each other.

fault-block mountain A mountain containing tall *horsts* interspersed with much lower *grabens* and bounded on at least one side by a high-angle *normal fault.*

faunal succession See *principle of faunal succession.*

firn Firmly packed snow that has survived a summer melting season. Firn has a density of about 0.4 grams per cubic centimeter. Ultimately, firn turns into glacial ice.

floodplain The flat land that surrounds a *stream* and becomes submerged when the stream overflows its banks.

flow A type of *mass movement* involving a mixture of solid, mostly unconsolidated particles that moves downslope like a viscous liquid although it may be relatively dry. The higher the water content of a flow, the more swiftly it moves.

focus (plural **foci**) The precise point within the Earth's *crust* or *mantle* where rocks begin to rupture or move in an *earthquake.*

fold A bend that develops in an initially horizontal layer of rock, usually caused by *plastic deformation.* Folds occur most frequently in *sedimentary rocks.*

fold-and-thrust mountains Large, complex mountain systems resulting from the collision of continental plates. Typically, these rocks consist of marine sediments that have been intensely *folded, thrust-faulted,* and, in places, intruded and metamorphosed by large *plutons.* The Alps, the Appalachians, the Carpathians, the Himalayas, and the Urals are all fold-and-thrust mountains.

foliation The arrangement of a set of minerals in parallel, sheet-like layers that lie perpendicular to the flattened plane of a rock. Occurs in *metamorphic rocks* on which *directed pressure* has been exerted.

foraminifera A group of protists that mostly build calcareous tests, includes benthic and planktonic forms.

formation A *stratigraphic* unit defined as a body of rock that is distinctive from the rock bodies above and below it, and large enough to be shown on a map.

fossil A remnant, an imprint, or a trace of an ancient organism, preserved in the Earth's crust.

fossil fuel A nonrenewable energy source, such as oil, gas, or coal, that derives from the organic remains of past life. Fossil fuels consist primarily of *hydrocarbons.*

fractional crystallization The process by which a *magma* produces crystals that then separate from the original magma, so that the chemical composition of the magma changes with each generation of crystals, producing *igneous rocks* of different compositions. The *silica* content of the magma becomes proportionately higher after each crystallization.

Franciscan Group Intensely-deformed oceanic and volcanic arc sediments and *ophiolites* that accumulated in the accretionary wedge formed by *subduction* of the *Farallon* plate during the *Jurassic* and *Cretaceous* periods.

fringing reef A reef that forms against or near an island or continental *coast* and grows seaward, sloping sharply toward the sea floor. Fringing reefs usually range from 0.5 to 1.0 or more kilometers in width.

frost wedging A form of *mechanical weathering* caused by the freezing of water that has entered a pore or crack in a rock. The water expands as it freezes, widening the cracks or pores and often loosening or dislodging rock fragments.

gabbro Any of a group of dark, dense, phaneritic, *intrusive rocks* that are the *plutonic* equivalent to *basalt.*

gastropods A *class* of *mollusks* characterized by a large fleshy foot used for locomotion. Most are shelled, including snails and whelks.

genes Inherited traits encoded by sections of the *DNA* sequence, defining the production of specific proteins that control metabolic functions in the cell.

genus The primary division of *family* in the Linnaean system of biological classification, consisting of one or more *species.*

geochronology The study of the relationship between the history of the Earth and time.

geologic time scale The division of all of Earth history into blocks of time distinguished by geologic and evolutionary events, ordered sequentially and arranged into *eons* made up of *eras,* which are in turn made up of *periods,* which are in turn made up of *epochs.*

geology The scientific study of the Earth, its origins and evolution, the materials that make it up, and the processes that act on it.

geyser A *natural spring* marked by the intermittent escape of hot water and steam.

gingkoes A group of *gymnosperms* common during the *Mesozoic* and surviving to the present.

glacial abrasion The process by which a *glacier* erodes the underlying bedrock through contact between the bedrock and rock fragments embedded in the base of the glacier. See also *glacial quarrying.*

glacial drift A load of rock material transported and deposited by a *glacier.* Glacial drift is usually deposited when the glacier begins to melt.

glacial erratic A rock or rock fragment transported by a *glacier* and deposited on bedrock of different composition. Glacial erratics range from a few millimeters to several yards in diameter.

glacial quarrying The process by which a *glacier* erodes the underlying bedrock by loosening and ultimately detaching blocks of rock from the bedrock and attaching them instead to the glacier, which then bears the rock fragments away. See also *glacial abrasion.*

glacial rebound Uplift in regions where the crust had been depressed by the weight of glaciers during the *Pleistocene Epoch.*

glacial till Drift that is deposited directly from glacial ice and therefore not *sorted.* Also called *till.* See also *glacial drift.*

glacier A moving body of ice that forms on land from the accumulation and compaction of snow, and that flows downslope or outward due to gravity and the pressure of its own weight.

gneiss A coarse-grained, foliated *metamorphic rock* marked by bands of light-colored minerals such as quartz and feldspar that alternate with bands of dark-colored minerals. This alternation develops through *metamorphic differentiation.*

Gondwana A *Paleozoic* continent encompassing the modern continents of South America, Africa, Australia, Antarctica, and India, and pieces of central and southeast Asia. Gondwana collided with *Laurasia* during the late Paleozoic to form Pangaea.

graben A block of rock that lies between two *faults* and has moved downward to form a depression between the two adjacent fault blocks. See also *horst.*

gracile australopithecines A group of smaller, mostly earlier *australopithecine species* including *A. anamensis, A. afarensis,* and *A. africanus.*

graded bed A *bed* formed by the deposition of sediment in relatively still water, marked by the presence of particles that vary in size, density, and shape. The particles settle in a gradual slope with the coarsest particles at the bottom and the finest at the top.

graded stream A stream maintaining an equilibrium between the processes of erosion and deposition, and therefore between aggradation and degradation.

gradient The vertical drop in a *stream's* elevation over a given horizontal distance, expressed as an angle.

granite A pink-colored, felsic, *plutonic rock* that contains potassium and usually sodium feldspars, and has a quartz content of about 10%. Granite is commonly found on continents but virtually absent from the ocean basins.

granulite gneiss One of the two primary rock associations in continental crust of Archean age, consisting of highly *metamorphosed igneous* rocks. Regions of granulite gneiss are separated by *greenstone belts* (see glossary definition).

graptolites Members of the hemichordate *phylum,* colonial organisms that often lived in floating clusters during the early to middle *Paleozoic.*

Great Valley Group *Sedimentary* rocks underlying the Great Valley of California that were deposited during the Jurassic and Cretaceous periods in the forearc basin that formed seaward of the Sierra Nevada magmatic arc.

Green River Formation Lake sediments deposited in the Green River, Uinta, and other basins during the Eocene, well known for the occurrence of fossils and oil shale.

greenstone belts Elongated belts of interlayered *volcanic* and *sedimentary rocks, metamorphosed* and folded into broad troughs. With *granulite gneisses* (see glossary definition), greenstone belts are one of the primary rock associations in continental *crust* of *Archean* age.

Grenville orogeny Tectonic event between 1.3 and 1.0 billion years ago during which the last major piece of the North American *craton* was added. This orogeny likely resulted from a substantial collision with another continent or series of smaller terranes.

groin A structure that juts out into a body of water perpendicular to the *shoreline* and is built to restore an eroding beach by intercepting *longshore drift* and trapping sand.

Gulf Stream Warm waters formed by the westward-flowing Atlantic current after the formation of the Isthmus of Panama between 3.5 and 3 million years ago.

gymnosperms Nonflowering, seed-bearing plants, including the conifers, first appearing during the late *Paleozoic.*

Hadean Eon The time before the *Archean Eon* for which there is little or no geologic record, lasting from 4.6 billion to 3.8 billion years ago.

half-life The time necessary for half of the atoms of a *parent isotope* to decay into the *daughter isotope.*

headland A cliff that projects out from a *coast* into deep water.

Hercynian orogeny A mountain-building event resulting from the late *Paleozoic* collision of the northern margin of *Gondwana,* consisting of southern Europe, with *Baltica* (northern Europe).

Himalayan orogeny Uplift of the Himalayas and Tibetan Plateau resulting from the Cenozoic collision of India and Eurasia.

historical geology The study of the origin and evolution of the Earth and all of its life forms and geologic structures.

Holocene Epoch The final epoch of the *Quaternary Period,* lasting from 10,000 years ago to the present.

hominids Hominoid *family* including extinct *australopithecines* and extant *genus Homo.*

Homo erectus Hominid species that lived from 1.8 million years ago to about 250,000 years ago, is the oldest hominid whose fossils are found outside of Africa.

homology The presence of similar skeletal structures that perform different functions in different organisms that share a common ancestry.

hook A *spit* that curves sharply at its coastal end due to a particularly strong *longshore current.*

horn A high mountain peak that forms when the walls of three or more *cirques* intersect.

hornfels A hard, dark-colored, dense *metamorphic rock* that forms from the *intrusion* of magma into shale or *basalt.*

horst A block of rock that lies between two *faults* and has moved upward relative to the two adjacent fault blocks. See also *graben.*

hot spot An area in the upper *mantle,* ranging from 100 to 200 km in width, from which magma rises in a plume to form *volcanoes.* A hot spot may endure for ten million years or more.

hydraulic lifting The *erosion* of a stream bed by water pressure.

hydrocarbon A molecule that is made up entirely of hydrogen and carbon.

hydrologic cycle The perpetual movement of water among the mantle, oceans, land, and atmosphere of the Earth.

hydrolysis A form of *chemical weathering* in which ions from water replace equivalently charged ions from a mineral, especially a silicate.

hydrothermal deposit A mineral deposit formed by the precipitation of metallic ions from water ranging in temperature from 50° to 700°C.

hypothesis A tentative explanation of a given set of data that is expected to remain valid after future observation and experimentation. See also *theory.*

Iapetus Ocean The ocean that separated *Laurentia* from *Baltica* during the early *Paleozoic Era* following the breakup of *Rodinia.*

ice age A period during which the Earth is substantially cooler than usual and a significant portion of its land surface is covered by *glaciers.* Ice ages generally last tens of millions of years.

icecap An *alpine glacier* that covers the peak of a mountain.

ichthyosaurs *Mesozoic* marine reptiles that had fishlike bodies and elongated snouts.

igneous rock A *rock* made from molten (melted) or partly molten material that has cooled and solidified.

inclusions See *principle of inclusions.*

index fossils Fossils that are particularly useful for *stratigraphic* correlation due to their short range, widespread occurrence, and ease of identification.

index mineral See *metamorphic index mineral.*

inheritance of acquired characteristics The theory of evolution proposed by the French naturalist Lamarck (1774–1829) who believed that organisms would develop certain useful traits that they would then pass on to their offspring. This theory was eventually disproved.

inorganic chert A type of *sedimentary rock* composed of microscopic silica crystals and formed by *precipitating* directly from silica-rich water. Typically found in the shape of nodules, often within bodies of limestone or *dolostone.* See also *biogenic chert.*

inorganic limestone A type of *sedimentary rock* composed mostly of calcite (calcium carbonate) and formed by *precipitating* directly from water. See also *biogenic limestone.*

insectivores Possibly the oldest order of placental mammals including modern shrews, moles, and hedgehogs.

inselberg A steep ridge or hill left in an otherwise flat, typically desert plain where a mountain has eroded.

internal deformation The rearrangement of the planes within ice crystals, due to pressure from overlying ice and snow, that causes the downward or outward flow of a *glacier.*

intracratonic basin An area of the *craton* that has subsided gradually over a long period of time allowing the accumulation of thick sequences of *sedimentary rocks.*

intrusive rock An *igneous rock* fromed by the entrance of *magma* into preexisting rock.

ion An *atom* that has lost or gained one or more *electrons,* thereby becoming electrically charged.

ionic bond The combination of an atom that has a strong tendency to lose electrons with an atom that has a strong tendency to gain electrons, such that the former transfers one or more electrons to the latter and each achieves chemical stability. The resulting *compound* is electrically neutral.

isotope One of two or more forms of a single element; the atoms of each isotope have the same number of protons but different numbers of neutrons in their nuclei. Thus, isotopes have the same *atomic number* but differ in *atomic mass.*

jetty A structure that is built to extend the banks of a stream channel or tidal outlet beyond the coastline to direct the flow of a stream or tide and keep the sediment moving so that it cannot build up and fill the channel. Jetties are typically built in parallel pairs along both banks of the channel. Jetties that are built perpendicular to a *coast* tend to interrupt *longshore drift* and thus widen *beaches.*

Juan de Fuca plate A remnant of the older *Farallon* ocean plate being subducted beneath northwestern North America.

Jurassic Period The second of the three time periods that make up the *Mesozoic Era* lasting from 205 million years ago until 140 million years ago.

karst The features that are produced, either at the Earth's surface or undergound (such as *caves*), when groundwater dissolves bedrock (usually limestone). See also *karst topography.*

karst topography The surface expression of *karst,* characterized principally by *sinkholes* and *disappearing streams.*

Kaskaskia sequence The third of the major *cratonic sequences* (see glossary definition) deposited during the *Paleozoic Era,* consisting of *Devonian* and *Mississippian* age *sediments* deposited by *transgression* and *regression* on the platform. The sequence is bounded above and below by *unconformities* in many places.

Kazakhstania A *Paleozoic* continent consisting of a portion of modern west-central Asia that collided with *Siberia* during the *Pennsylvanian Period.*

kerogen A solid, waxy, organic substance that forms when pressure and heat from the Earth act on the remains of plants and animals. Kerogen converts to various liquid and gaseous *hydrocarbons* at a depth of 7 or more kilometers and a temperature between 50° and 100°C.

kingdom The primary division of life in the Linnaean system of biological classification. Four kingdoms of eukaryotes are now recognized and the animal kingdom is divided into many *phyla.*

komatiites Ultramafic (Fe-, Mg-rich) volcanic rocks that formed during the *Archean* when the Earth was hotter, most commonly found in *greenstone belts* (see glossary definition).

labyrinthodonts A diverse group of late *Paleozoic* amphibians characterized by a convoluted tooth structure.

laccolith A large *concordant pluton* that is shaped like a dome or a mushroom. Laccoliths tend to form at relatively shallow depths and are typically composed of granite. The country rock above them often erodes away completely.

lagomorphs An order of gnawing mammals, including rabbits and hares, closely related to *rodents.*

lahar A flow of pyroclastic material mixed with water. A lahar is often produced when a snow-capped volcano erupts and hot pyroclastics melt a large amount of snow or ice.

Laramide orogeny Late *Cretaceous* to Early *Tertiary* mountain-building episode in the *Cordillera* characterized by forming broad uplifts and steep-sided folds located to the east of the *Sevier* thrust belt.

Laurasia A large continent formed during the middle *Paleozoic* by the collision and suturing of *Laurentia* and *Baltica* (the *Caledonian orogeny*–see glossary definition).

Laurentia A *Proterozoic* to early *Paleozoic* continent which included North America and parts of Greenland, northern Great Britain, and Scandinavia.

Laurentide ice sheet A continental glacier that extended southward from northern Canada during the *Pleistocene Epoch.*

lava *Magma* that comes to the Earth's surface through a *volcano* or fissure.

lignite A soft, brownish coal that develops from *peat* through bacterial action, is rich in *kerogen,* and has a carbon content of 70%, which makes it a more efficient heating fuel than peat.

liquefaction The conversion of moderately cohesive, unconsolidated *sediment* into a fluid, water-saturated mass.

lithification The conversion of loose *sediment* into solid *sedimentary rock.*

lithosphere A layer of solid, brittle rock comprising the outer 100 kilometers of the Earth, encompassing both the crust and the outermost part of the upper *mantle.* See also *asthenosphere.*

lithostatic pressure The force exerted on a rock buried deep within the Earth by overlying rocks. Because lithostatic pressure is exerted equally from all sides of a rock, it compresses the rock into a smaller, denser form without altering the rock's shape.

lobe-finned fish A group of bony fish in which the fins form a muscular limb, includes the lungfish.

loess A blanket of silt that is produced by the erosion of glacial outwash and transported by wind. Much loess found in the Mississippi Valley, China, and Europe is believed to have been deposited during the *Pleistocene Epoch.*

longitudinal dune One of a series of long, narrow *dunes* lying parallel both to each other and to the prevailing wind direction.

longshore current An ocean current that flows close and almost parallel to the *shoreline,* caused by the combination of swash and backwash.

longshore drift 1. The process by which a *longshore current* moves *sediments* within a surf zone. 2. The sediments so moved. Longshore drift typically consists of sand, gravel, shell fragments, and pebbles. See also *beach drift.*

lopolith A large, saucer-shaped concordant *pluton* produced when dense mafic magma depresses the country rock below it.

lycopsids A group of *seedless vascular plants* (see glossary definition) that grew large trees during the *Paleozoic Era,* now represented by the ground pine.

magma Molten (melted) rock that forms naturally within the Earth. Magma may be either a liquid or a fluid mixture of liquid, crystals, and dissolved gases.

mantle The middle layer of the Earth, lying just below the *crust* and consisting of relatively dense rocks. The mantle is divided into two sections, the upper mantle and the lower mantle; the lower mantle has greater density than the upper mantle. See also *core* and *crust.*

marble A coarse-grained, nonfoliated *metamorphic rock* derived from limestone or *dolostone.*

marine magnetic anomaly An irregularity in magnetic strength along the ocean floor that reflects *sea-floor spreading* during periods of magnetic reversal.

marsupials Mammals that give birth to embryonic young that then continue to develop in an external pouch.

mass extinction The extinction of a large percentage of the existing *species, genera, families,* or *orders* within a geologically short period of time.

mass movement The process by which such Earth materials as bedrock, loose *sediment,* and *soil* are transported down slopes by gravity.

meandering stream A *stream* that traverses relatively flat land in fairly evenly spaced loops and separated from each other by narrow strips of *floodplain.*

mechanical exfoliation A form of *mechanical weathering* in which successive layers of a large *plutonic rock* break loose and fall when the erosion of overlying material permits the rock to expand upward. The thin slabs of rock that break off fall parallel to the exposed surface of the rock, creating the long, broad steps that can be found on many mountains.

mechanical weathering The process by which a rock or mineral is broken down into smaller fragments without altering its chemical makeup; *weathering* that affects only physical characteristics. See also *chemical weathering.*

meiosis The process of formation of gametes for sexual reproduction, involves duplication of the chromosomes followed by two stages of cell division, resulting in reproductive cells (eggs and sperm) with a one-half complement of chromosomes.

Mercalli intensity scale A scale designed to measure the degree of intensity of *earthquakes,* ranging from I for the lowest intensity to XII for the highest. The classifications are based on human perceptions.

Mesozoic Era The intermediate era of the *Phanerozoic Eon,* following the *Paleozoic Era* and preceding the *Cenozoic Era,* and marked by the dominance of marine and terrestrial reptiles and the appearance of birds, mammals, and flowering plants.

metallic bonding The act or process by which two or more atoms of electron-donating elements pack so closely together that some of their electrons begin to wander among the nuclei rather than orbiting the nucleus of a single atom. Metallic bonding is responsible for the distinctive properties of metals.

metamorphic differentiation The process by which minerals from a chemically uniform rock separate from each other during *metamorphism* and form individual layers within a new *metamorphic rock.*

metamorphic facies 1. A group of minerals customarily found together in *metamorphic rocks* and indicating a particular set of temperature and pressure conditions at which metamorphism occurred. 2. A set of *metamorphic rocks* characterized by the presence of such a group of minerals.

metamorphic grade A measure used to identify the degree to which a *metamorphic rock* has changed from its parent rock. A metamorphic grade provides some indication of the circumstances under which the metamorphism took place.

metamorphic index mineral One of a set of minerals found in *metamorphic rocks* and used as indicators of the temperature and pressure conditions at which the metamorphism occurred. A metamorphic index mineral is stable only within a narrow range of temperatures and pressures and the metamorphism that produces it must take place within that range.

metamorphic rock A *rock* that has undergone chemical or structural changes. Heat, pressure, or a chemical reaction may cause such changes.

metamorphism The process by which conditions within the Earth alter the mineral content, chemical composition, and structure of solid rock without melting it. *Igneous, sedimentary,* and *metamorphic rocks* may all undergo metamorphism.

microcontinent A section of continental *lithosphere* that has broken off from a large, distant continent, as by *rifting.*

Midcontinent rift A string of *normal faults* and down-dropped *basins* filled with *lavas* and terrestrial *sediments* extending 1500 kilometers across central North America, formed by stretching of the *crust* during the Middle *Proterozoic.*

mid-ocean ridge An underwater mountain range that develops between the margins of diverging plates, produced by accumulation of mantle basalt that continues to erupt after plate *rifting.*

migmatite A rock that incoporates both *metamorphic* and *igneous* materials.

mineral A naturally occurring, usually inorganic, solid consisting of either a single element or a *compound,* and having a definite chemical composition and a systematic internal arrangement of atoms.

mineraloid A naturally occurring, usually inorganic, solid consisting of either a single element or a *compound,* and having a definite chemical composition but lacking a systemic internal arrangement of atoms. See also *mineral.*

mineral zone An area of rock throughout which a given *metamorphic index mineral* is found, presumed to have undergone metamorphism under uniform temperature and pressure conditions.

Miocene Epoch An epoch of the *Tertiary Period,* lasting from 24 million years ago to 5 million years ago.

Mississippian Period A period of late *Paleozoic* time defined in the United States as lasting from 360 million years ago to 320 million years ago; corresponds to the Lower *Carboniferous* used elsewhere.

mitosis Process of cell division involving duplication of the chromosomes; most single-celled organisms reproduce this way.

Moho (abbreviation for Mohorovičić) The seismic discontinuity between the base of the Earth's *crust* and the top of the *mantle. P waves* passing through the Moho change their velocity by approximately 1 kilometer per second, with the higher velocity occurring in the mantle and the lower in the crust.

mollusks A diverse invertebrate *phylum* including *bivalves, gastropods,* and *cephalopods.*

monotremes Primitive mammals that lay eggs but produce milk to nourish their young, first appearing in the *Cretaceous* and including modern platypus and spiny anteaters.

moraine A single, large mass of *glacial till* that accumulates, typically at the edge of a glacier.

Morrison Formation A *formation* (see glossary definition) consisting of *alluvial sediments* deposited in the western foreland basin during the Late *Jurassic,* famous for *dinosaur* fossils.

mosasaurs A group of large marine lizards that lived during the *Cretaceous.*

mudcrack A fracture that develops at the top of a layer of fine-grained, muddy sediment when it is exposed to the air and as a result dries out and contracts.

mudflow The rapid flow of typically fine-grained *regolith* mixed with water. There may be as much as 60% water in a mudflow.

mudstones Fine-grained *sedimentary rocks* composed largely of silt, clay minerals, and mica that have settled out of still water. These include shale and siltstone.

multiregional origin The theory that modern humans evolved simultaneously in several locations from a common ancestral group. Contrasts with the "*out of Africa*" theory (see glossary definition).

mutation A change in the genetic sequence caused by environmental factors or mistakes in *DNA* copying that may give rise to a new trait, beneficial or harmful, in an organism.

native elements *Elements* that do not combine with others in nature. Minerals composed of native elements, such as gold, silver, and diamond, contain only those elements.

natural levee One of a pair of ridges of *sediment* deposited along both banks of a *stream* during successive floods.

natural selection The mechanism of evolution as proposed by Darwin in which competition for food and mates would select individuals with certain favorable traits for survival. These individuals would pass those traits on to subsequent generations.

natural spring A place where groundwater flows to the surface and issues freely from the ground.

Navajo Sandstone A *formation* (see glossary definition) consisting largely of wind-blown sand dunes deposited in the West during the Early *Jurassic.*

Neanderthal Hominids with muscular bodies and large brains that lived in Europe and the Near East from 200,000 years ago to 30,000 years ago, considered a separate human *species* by some researchers, or a subspecies of *H. sapiens* by others.

neutron A particle that is found in the *nucleus* of an *atom,* has a mass approximately equal to that of a *proton,* and has no electric charge.

Nevadan orogeny An interval of intensified *igneous* activity during the Late *Jurassic* to Early *Cretaceous* caused by *subduction* of the *Farallon plate* beneath North America, resulting in the formation of the major igneous *batholiths* (see glossary definition) of the West.

Newark Supergroup Collective term for the *sedimentary* and *volcanic rocks* that accumulated in all of the *Mesozoic rift basins* of eastern North America.

normal fault A *dip-slip fault* marked by a generally steep *dip* along which the hanging wall has moved downward relative to the footwall.

nuclear fission The division of the *nuclei* of *isotopes* of certain heavy *elements,* such as uranium and plutonium, effected by bombardment with *neutrons.* Nuclear fission causes the release of energy, additional neutrons, and an enormous quantity of heat.

nuclear fusion The combination of the *nuclei* of certain extremely light *elements,* especially hydrogen, effected by the application of high temperature and pressure. Nuclear fusion causes the release of an enormous amount of heat energy, comparable to that released by *nuclear fission.*

nucleus (plural **nuclei**) 1. The central part of an *atom,* containing most of the atom's mass and having a positive charge due to the presence of *protons.* 2. The membrane-bound structure found in the cells of *eukaryotic* organisms that contains their genetic information, in the form of DNA.

nuée ardente A sometimes glowing cloud of gas and *pyroclastics* erupted from a *volcano* and moving swiftly down its slopes. Also called a *pyroclastic flow.*

ocean trench A deep, linear, relatively narrow depression in the sea floor, formed by the *subduction* of oceanic plates.

odd-toed ungulates Group of hoofed mammals that includes modern horses, rhinoceroses, and tapirs.

oil sand A mixture of unconsolidated sand and clay that contains *bitumen,* a semi-solid *hydrocarbon.*

oil shale A brown or black clastic *source rock* containing *kerogen.*

Oligocene Epoch An epoch of the middle *Tertiary Period,* lasting from 38 million years ago to 24 million years ago.

ophiolite suite The group of *sediments, sedimentary rocks,* and mafic and ultramafic *igneous rocks* that make up the oceanic *lithosphere.*

order The primary division of *class* (see glossary definition) in the Linnaean system of biological classification, consisting of one or more *families.*

Ordovician Period A period of early *Paleozoic* time lasting from 505 million years ago until 438 million years ago.

ore A mineral deposit that can be mined for a profit.

original horizontality See *principle of original horizontality.*

ornithischians The "bird-hipped" dinosaurs in which the pubis and ischium point backward, as in modern birds. Includes the *stegosaurs, ceratopsians, ankylasaurs, ornithopods,* and the *pachycephalosaurs.*

ornithopods The "duck-billed" dinosaurs, a group of bipedal *ornithischians* distinguished by flattened snouts.

orogen An elongated belt of deformed rocks formed at the edge of a *convergent* plate margin.

orogenesis Mountain formation, as caused by *volcanism, subduction, plate divergence, folding,* or the movement of fault blocks. Also called *orogeny.*

oscillatory motion The circular movement of water up and down, with little or no change in position, as a wave passes.

ostracoderms The earliest recognized fish, an extinct group of jawless fish, characterized by platy skin, that lived during the *Paleozoic Era.*

Ouachita orogeny Mountain-building event during the late *Paleozoic Era* resulting from the collision of the South American margin of *Gondwana* and the southern margin of North America that formed an elongated *fold-and-thrust mountain* belt (see glossary definition) exposed in the Ouachita Mountains of Oklahoma and Arkansas.

outgassing The release of gases and water vapor to the Earth's surface by *volcanic* activity, responsible for the formation of the first permanent atmosphere.

out of Africa The theory that modern humans evolved in Africa between 100,000 and 300,000 years ago and later migrated out of Africa to populate the world. Contrasts with the "*multiregional origin*" theory (see glossary definition).

outwash A load of *sediment,* consisting of sand and gravel, that is deposited by meltwater in front of a *glacier.*

overkill hypothesis An explanation for the late Quaternary extinctions, attributing overhunting by humans for the loss of large land mammals.

oxidation The process whereby a mineral's ions combine with oxygen ions. A mineral that is exposed to air may undergo oxidation as a form of *chemical weathering.*

oxide One of seveal minerals containing negative oxygen ions bonded to one or more positive metallic ions.

pachycephalosaurs A group of bipedal *ornithischians* distinguished by a dome-like thickening of the skull.

Paleocene Epoch The first epoch of the *Tertiary Period,* lasting from 66 million years ago to 58 million years ago.

paleoclimate The climate of the past.

paleoecology The relationship between organisms and the ancient environment in which they lived.

paleogeography The geographic arrangement of landmasses, land features, and bodies of water for periods in the past.

Paleozoic Era The earliest era of the *Phanerozoic Eon,* marked by the presence of marine invertebrates, fish, amphibians, insects, and land plants.

Panthalassa The ocean that surrounded the supercontinent Pangaea from the late *Paleozoic* until breakup during the *Mesozoic.*

parabolic dune A horseshoe-shaped *dune* having a concave windward slope and a convex leeward slope. Parabolic dunes tend to form along sandy ocean and lake shores. They may also develop from *transverse dunes* through *deflation.*

parent isotope A radioactive *isotope* that changes into a different isotope when its nucleus decays. See also *daughter isotope.*

parent material The source from which a given soil is chiefly derived, generally consisting of bedrock or *sediment.*

partial melting The incomplete melting of a rock composed of minerals with differing melting points. When partial melting occurs, the minerals with higher melting points remain solid while the minerals whose melting points have been reached turn into *magma.*

passive continental margin A border that lies between continental and oceanic *lithosphere,* but is not a plate margin. It is marked by lack of seismic and volcanic activity.

peat A soft, brown mass of compressed, partially decomposed vegetation that forms in a water-saturated environment and has a carbon content of 50%. Dried peat can be burned as fuel.

pediment A broad surface at the base of a receding mountain. The pediment develops when running water erodes most of the mass of the mountain.

pelycosaurs A group of Permian reptiles, most of which had prominent sails on their backs, including carnivorous and herbivorous forms.

Pennsylvanian Period A period of late *Paleozoic* time defined in the United States lasting from 320 million years ago until 286 million years ago, corresponding to the Upper *Carboniferous Period* (see glossary definition) used elsewhere.

peridotite An *igneous rock* composed primarily of the iron-magnesium *silicate* olivine and having a silica content of less than 40%.

periods The time spans that subdivide *eras* in the *geologic time scale,* and are themselves divided into *epochs.*

permeability The capability of a given substance to allow the passage of a fluid. Permeability depends upon the size of and the degree of connection among a substance's pores.

Permian Period The final period of the *Paleozoic Era,* lasting from 286 million years ago until 245 million years ago.

petroleum The most common and versatile fossil fuel, comprising a group of naturally occurring substances made up of *hydrocarbons.* These substances may be gaseous, liquid, or semi-solid.

Phanerozoic Eon The eon that started 544 million years ago, when numerous fossils of sea shells began to be formed, and that continues to the present time.

photochemical dissociation The splitting of water molecules by ultraviolet radiation, forming hydrogen, which is lost to space, and free oxygen. A relatively minor amount of atmospheric oxygen was formed this way.

phyletic gradualism The view that evolution progresses through small changes that are passed through the entire population over many generations, gradually transforming ancestral *species* into new ones.

phyllite A foliated *metamorphic rock* that develops from *slate* and is marked by a silky sheen and medium grain size.

phylum The primary division of *kingdom* (see glossary definition) in the Linnaean system of biological classification, consisting of one or more *classes.*

placental Mammals in which the young develop within the uterus prior to birth, first appearing during the Cretaceous, includes most modern mammals.

placer deposit A deposit of heavy or durable minerals, such as gold or diamonds, typically found where the flow of water abruptly slows.

placoderms A *class* of extinct armored fish with jaws that lived during the *Paleozoic Era,* some of which grew to impressive size.

plastic deformation A permanent change in the shape or volume of a rock caused by great *stress* under higher temperatures and pressures than that which produce *brittle failure.*

plate tectonics The theory that the Earth's *lithosphere* consists of large, rigid plates that move horizontally in response to the flow of the *asthenosphere* beneath them, and that interactions among the plates at their borders cause most major geologic activity, including the creation of oceans, continents, mountains, volcanoes, and earthquakes.

playa A dry lake basin found in a desert.

Pleistocene Epoch The first epoch of the *Quaternary Period,* lasting from 1.6 million years ago to 10,000 years ago, largely coinciding with the last ice age.

plesiosaurs Extinct *Mesozoic* marine reptiles with flipperlike limbs and slender necks.

Pliocene Epoch The final epoch of the *Tertiary Period,* lasting from 5 million years ago to 1.6 million years ago.

pluton An *intrusive rock,* as distinguished from the preexisting country rock that surrounds it.

plutonic rock See *intrusive rock.*

pluvial lakes Lakes created by abundant rainfall in basins of the desert Southwest during the *Pleistocene Epoch.*

point bar A low ridge of *sediment* that forms along the inner bank of a *meandering stream.*

polymorph A mineral that is identical to another mineral in chemical composition but differs from it in *crystal structure.*

porosity The percentage of a soil, rock, or sediment's volume that is made up of pores.

primary coast A *coast* formed primarily by nonmarine processes, such as *glacial erosion,* sea-level fluctuations, or biological processes. See also *secondary coast.*

primates Placental mammals with five digits on the extremities, which are adapted for grasping, includes *prosimians* and *anthropoids* (see glossary definitions).

principle of cross-cutting relationships The scientific law stating that a *pluton* is always younger than the rock that surrounds it.

principle of faunal succession The scientific law stating that the organisms of Earth have evolved in a definite order over time, and that this is reflected in the fossil record.

principle of inclusions The scientific law stating that rock fragments contained within a larger body of rock are always older than the surrounding body of rock.

principle of original horizontality The scientific law stating that *sediments* setting out from bodies of water are deposited horizontally or nearly horizontally in layers that lie parallel or nearly parallel to the Earth's surface.

principle of superposition The scientific law stating that in any unaltered sequence of rock strata, each stratum is younger than the one beneath it and older than the one above it, so that the youngest stratum will be at the top of the sequence and the oldest at the bottom.

principle of uniformitarianism The scientific law stating that the geological processes taking place in the present operated similarly in the past and can therefore be used to explain past geologic events.

proboscideans Large land mammals most of which are distinguished by trunks and tusks, includes extinct mammoths and mastodons and modern elephants.

prokaryotic Lacking a cell *nucleus.* Describes the first simple one-celled organisms to appear on Earth as well as modern bacteria and archaea. See also *eukaryotic.*

prosimians Primitive *primates* that are mostly smaller climbing creatures, includes the tree shrews, lemurs, lorises, and tarsiers.

Proterozoic Eon The final division of the Precambrian, lasting from 2.5 billion years ago until 544 million years ago.

proton A positively charged particle that is found in the *nucleus* of an *atom* and has a mass approximately 1836 times that of an *electron.*

pterosaurs A group of extinct flying reptiles that lived during the *Mesozoic.*

punctuated equilibrium The view that large populations of an individual *species* (see glossary definition) tend to remain unchanged for extended periods of time while geographically isolated groups may undergo rapid change and give rise to new species which may quickly replace the original.

P wave (abbreviation for **primary wave**) A *body wave* that causes the *compression* of rocks when its energy acts upon them. When the P wave moves past a rock, the rock expands beyond its original volume, only to be compressed again by the next P wave. P waves are the fastest of all *seismic waves.* See also *S wave.*

pyroclastic cone A usually steep, conic *volcano* composed almost entirely of an accumulation of loose pyroclastic material. Pyroclastic cones are usually less than 450 meters high. Because no *lava* binds the *pyroclastics,* pyroclastic cones erode easily.

pyroclastic eruption A volcanic eruption of viscous, gas-rich magma. Pyroclastic eruptions tend to produce a great deal of solid volcanic fragments rather than fluid *lava.*

pyroclastic flow A rapid, extremely hot, downward stream of *pyroclastics,* air, gases, and ash ejected from an erupting *volcano.* A pyroclastic flow may be as hot as 800°C or more and may move at speeds higher than 150 kilometers per hour.

pyroclastics (used only in the plural) Particles and chunks of *igneous rock* ejected from a volcanic vent during an eruption.

quake See *earthquake.*

quartzite An extremely durable, nonfoliated *metamorphic rock* derived from pure *sandstone* and consisting primarily of quartz.

Quaternary Period The second period of the *Cenozoic Era,* lasting from 1.6 million years ago to the present. Subdivided into the *Pleistocene* and *Holocene* epochs.

quick clay A partly waterlogged solid clay sediment that almost instantaneously becomes a highly fluid *mudflow,* usually when ground vibrations increase the water pressure between its particles, causing them to separate and reducing the friction between them.

radiometric dating The process of using relative proportions of *parent* to *daughter isotopes* in radioactive decay to determine the age of a given rock or rock stratum.

rain-shadow desert A desert that forms when moist air on the windward side of a mountain rises and cools, causing precipitation there and leaving the leeward side of the mountain dry.

range The length of time an individual *species, genus,* or *family* exists.

ray-finned fish The largest group of bony fish, possessing fins in which the bones radiate away from the body, includes most familiar modern fish.

recrystallization The process by which unstable minerals in buried sediment are transformed into stable ones.

redbeds *Alluvial sediments,* typically shales and *sandstones,* with a red-brown color from the presence of oxidized iron minerals, such as hematite.

regional metamorphism *Metamorphism* that affects rocks over vast geographic areas stretching for thousands of square kilometers.

regolith The unconsolidated products of *mechanical* and *chemical weathering* that cover almost all of the Earth's land surface; composed of *soil, sediment,* and fragments of underlying bedrock.

regression The migration of the shoreline away from the continent, either from eustatic lowering of sea level, regional tectonic uplift, or deposition of shoreline *sediments.*

relative dating The fixing of a geologic structure or event in a chronological sequence relative to other geologic structures or events. See also *absolute dating.*

reserve A known *resource* that can be exploited for profit with available technology under existing political and economic conditions.

reservoir rock A permeable rock containing oil or gas.

resource A mineral or fuel deposit, known or not yet discovered, that is not currently available for human exploitation.

reverse fault A *dip-slip fault* marked by a hanging wall that has moved upward relative to the footwall. Reverse faults are often caused by the *convergence* of lithospheric plates.

rhyolite Any of a group of felsic *igneous rocks* that are the *extrusive* equivalents of *granite.*

Richter scale A logarithmic scale that measures the amount of energy released during an *earthquake* on the basis of the amplitude of the highest peak recorded on a seismogram. Each unit increase in the Richter scale represents a 10-fold increase in the amplitude recorded on the seismogram and a 30-fold increase in energy released by the earthquake.

rifting The tearing apart of a *plate* to form a depression in the Earth's *crust* and often eventually separating the plate into two or more smaller plates.

rip current A strong, rapid, and brief current that flows out to sea, moving perpendicular to the *shoreline.*

ripple marks A pattern of wavy lines formed along the top of a bed by wind, water currents, or waves.

robust australopithecines A group of larger, mostly later *australopithecine species* that includes *A. boisei* and *A. robustus.* Considered a separate *genus* (*Paranthropus*) by some.

rock A naturally formed aggregate of usually inorganic materials from within the Earth.

rock-forming mineral One of the twenty or so minerals contained in the rocks that compose the Earth's crust and mantle.

rodents The order of placental mammals with prominent, continuously growing incisor teeth, consisting of gnawing animals, such as mice, rats, and hamsters.

Rodinia A supercontinent that formed during the Middle *Proterozoic* and broke apart during the Late Proterozoic.

rudists *Mesozoic bivalves,* with one large horn-shaped lower valve and a smaller upper valve, were reef-formers during the *Cretaceous.*

rugose corals An extinct group of corals that lived during the *Paleozoic Era,* characterized by horn-shaped skeletons.

ruminants *Even-toed ungulates* characterized by a multi-chambered stomach; includes sheep, cattle, deer, antelope, and goats.

saltwater intrusion Infiltration of a coastal *aquifer* by salty marine water, which may occur if the aquifer is depleted of fresh water from overuse or lack of *precipitation.*

San Andreas transform A transform fault that connects the Juan de Fuca ridge and the East Pacific Rise, cutting across the western edge of North America.

sandstone A *clastic rock* composed of particles that range in diameter from 1/16 millimeter to 2 millimeters. Sandstones make up about 25% of all sedimentary rocks.

saturation zone See *zone of saturation.*

Sauk sequence The first of the major *cratonic sequences* (see glossary definition) deposited during the *Paleozoic Era,* consisting of *Cambrian* and *Ordovician* age *sediments* deposited by *transgression* and *regression* on the platform. The sequence is bounded above and below by *unconformities* in many places.

saurischians The "lizard-hipped" dinosaurs, in which the three hip bones point in different directions, includes the *theropods* and the *sauropods.*

sauropods Giant, quadrupedal *saurischian* dinosaurs that were plant-eaters, exemplified by *Apatosaurus.*

scarp The steep cliff face that is formed by a *slump.*

schist A coarse-grained, strongly foliated *metamorphic rock* that develops from *phyllite* and splits easily into flat, parallel slabs.

scientific law 1. A natural phenomenon that has been proven to occur invariably whenever certain conditions are met. 2. A formal statement describing such a phenomenon and the conditions under which it occurs. Also called *law.*

scientific method The technique that involves gathering all available data on a subject, forming a *hypothesis* to explain the data, conducting experiments to test the hypothesis, and modifying or confirming the hypothesis as necessary to account for the experimental results.

scleractinian coral The *order* of corals including all modern reef-forming varieties, first appearing during the Mesozoic.

sea arch A span of rock produced when *sea caves* on both sides of a coastal cliff, or *headland,* erode toward one another until they are joined.

sea cave A cave eroded out of a coastal cliff, or *headland,* by the battering of waves refracted against its flanks. (See *wave refraction.*)

sea-floor spreading The formation and growth of oceans that occurs following *rifting* and is characterized by eruptions along *mid-ocean ridges,* forming new oceanic *lithosphere,* and expanding ocean basins. See also *divergence.*

sea stack A steep, isolated island of rock, separated from a *headland* by the action of waves, as when the overhanging section of a *sea arch* is eroded.

secondary coast A *coast* shaped primarily by ongoing marine erosion or deposition, such as by sea currents and waves. See also *primary coast.*

sediment A collection of transported fragments or precipitated materials that accumulate, typically in loose layers, as of sand or mud.

sedimentary environment The continental, oceanic, or coastal surroundings in which sediment accumulates.

sedimentary facies 1. A set of characteristics that distinguish a given section of sedimentary rock from nearby sections. Such characteristics include mineral content, grain size, shape, and density. 2. A section of sedimentary rock so characterized.

sedimentary rock A *rock* made from the consolidation of solid fragments, as of other rocks or organic remains, or by *precipitation* of minerals from solution.

sedimentary structure A physical characterisic of a *detrital sediment* that reflects the conditions under which the sediment was deposited.

seedless vascular plants All spore-bearing plants with vascular tissues, including ferns, *lycopsids,* and *sphenopsids.* They were dominant in middle to late *Paleozoic* swamps.

seiche The movement of waves back and forth in an enclosed or partially enclosed basin, such as a lake or bay, due to *seismic waves* during an *earthquake.*

seismic gap A locked *fault* segment that has not experienced seismic activity for a long time. Because *stress* tends to accumulate in seismic gaps, they often become the sites of major *earthquakes.*

seismic wave One of a series of progressive disturbances that reverberate through the Earth of transmit the energy released from an *earthquake.*

Sevier orogeny A *Cretaceous* mountain-building event in the *Cordillera* during which slices of *sedimentary rock* from the foreland basin were *folded* and *thrust* eastward.

shearing stress *Stress* that slices rocks into parallel blocks that slide in opposite directions along their adjacent sides. Shearing stress may be caused by transform motion.

shield volcano A low, broad, gently sloping, dome-shaped structure that forms over time as repeated eruptions eject *basaltic lava* through one or more vents and the lava solidifies in approximately the same volume all around.

shocked quartz Grains of quartz containing multiple sets of parallel fractures, a feature formed by high impacts.

shoreline The boundary between a body of water and dry land.

Siberia A continent which included most of northern Asia (principally eastern Russia) that existed independently following the breakup of *Rodinia* until the late *Paleozoic.*

silicate One of several rock-forming minerals that contain silicon, oxygen, and usually one or more other common elements.

silicon-oxygen tetrahedron A four-sided geometric form created by the tight bonding of four oxygen atoms to each other, and also to a single silicon atom that lies in the middle of the form.

sill A concordant tabular *pluton* that is substantially wider than it is thick. Sills form within a few kilometers of the Earth's surface. See also *dike.*

Silurian Period A period of early *Paleozoic* time lasting from 438 million years ago until 408 million years ago.

sinkhole A circular, often funnel-shaped depression in the ground that forms when soluble rocks dissolve.

slate A fine-grained, foliated *metamorphic rock* that develops from shale and tends to break into thin, flat sheets.

slide The *mass movement* of a single, intact mass of rock, soil, or unconsolidated matrerial along a weak plane, such as a *fault,* fracture, or *bedding* plane. A slide may involve as little as a minor displacement of soil or as much as the displacement of an entire mountainside.

slump 1. A downward and outward *slide* occurring along a concave slip plane. 2. The material that breaks off in such a slide.

small, shelly fauna A group of tiny animals with calcareous shells, thought to be early representatives of sponges, *mollusks,* and *brachiopods,* that lived during the earliest *Cambrian Period.*

soil The top few meters of *regolith,* containing both mineral and organic matter.

soil horizon A layer of soil that can be distinguished from the surrounding soil by such features as chemical composition, color, and texture.

soil profile A vertical strip of soil stretching from the surface down to the bedrock and including all of the successive *soil horizons* in a given location.

soil taxonomy A soil calssification system that categorizes soils based on such attributes as their physical characteristics, chemistry, origin, and relative age.

solifluction A form of *creep* in which soil flows downslope at 0.5 to 15 centimeters per year. Solifluction occurs in relatively cold regions when the brief warmth of summer thaws only the upper meter or two of *regolith,* which becomes waterlogged because the underlying ground remains frozen and therefore the water cannot drain down into it.

Sonoma orogeny A mountain-building event in the *Cordillera* resulting from the collision and accretion of *Sonomia* (see glossary definition) during the *Permian* to *Triassic.*

Sonomia A *volcanic arc* terrane accreted to North America during the *Sonoma orogeny* at the end of the *Permian* and beginning of the *Triassic* periods.

sorting The process by which a given transport medium separates out certain particles, as on the basis of size, shape, or density.

source rock A rock in which *hydrocarbons* originate.

speciation The appearance of a new *species* (see glossary definition).

species The most basic unit of biological classification, defined as a population of organisms with similar appearance that interbreeds to produce fertile offspring.

speleothem A mineral deposit of calcium carbonate that precipitates from solution in a *cave.*

sphenopsids A group of *seedless vascular plants* (see glossary definition) that grew large trees during the *Paleozoic Era,* now represented by the horsetail.

spheroidal weathering The process by which *chemical weathering,* especially by water, decomposes the angles and edges of a rock or boulder, leaving a rounded form from which concentric layers are then stripped away as the weathering continues.

spit A narrow, finger-like ridge of sand that extends from land into the open water of a coastal bay, deposited by a *longshore current* interrupted by deeper water.

star dune A *dune* with three or four arms radiating from its usually higher center so that it resembles a star in shape. Star dunes form when winds blow from three or four directions, or when the wind direction shifts frequently.

stegosaurs A group of quadrupedal *ornithischians* characterized by a row of bony plates rising from the back.

strain The change in the shape or volume of a rock that results from *stress.*

stratification See *bedding.*

stratigraphy The arrangement of rock layers or the field of study of sequences of *sedimentary rocks.*

stratovolcano A cone-shaped *volcano* built from alternating layers of *pyroclastics* and viscous *andesitic lava.* Stratovolcanoes tend to be very large and steep.

stream A body of water found on the Earth's surface and confined to a narrow topographic depression, or channel, down which it flows and transports rock particles, sediment, and dissolved particles. Rivers, creeks, brooks, and runs are all streams.

stress The force acting on a rock or another solid to deform it, measured in kilograms per square centimeter or pounds per square inch.

strike 1. The horizontal line marking the intersection between the inclined plane of a solid geological structure and the Earth's surface. 2. The compass direction of this line, measured in degrees from true north.

strike-slip fault A *fault* in which two sections of rock have moved horizontally in opposite directions, parallel to the line of the fracture that divided them. Strike-slip faults are caused by *shearing stress.*

stromatolites Finely layered, moundlike structures of limestone built by mats of photosynthetic bacteria or algae living in very shallow marine or lake environments.

stromatoporoids An extinct group of calcareous sponges that were important reef-building organisms during the *Paleozoic Era.*

subduction The sinking of an oceanic *plate* edge as a result of *convergence* with a plate of lesser density. Subduction often causes *earthquakes* and creates *volcano* chains.

subsidence The lowering of the Earth's surface, caused by such factors as compaction, a decrease in groundwater, or the pumping of oil.

sulfate One of several minerals containing positive sulfur ions bonded to negative oxygen ions.

sulfide One of several minerals containing negative sulfur ions bonded to one or more positive metallic ions.

Sundance Sea An *epeiric sea* (see glossary definition) that covered much of the western *craton* resulting from the Zuni *transgression* during the *Jurassic Period.*

superposition See *principle of superposition.*

surface wave One of a series of *seismic waves* that transmits energy from an earthquake's *epicenter* along the Earth's surface. See also *body wave.*

suspended load A body of fine, solid particles, typically of sand, clay, and silt, that travels with stream water without coming into contact with the stream bed.

suture zone An area where two continental plates have joined together through *continental collision.* Suture zones are marked by extremely high mountain ranges, such as the Himalayas and the Alps.

S wave (abbreviation for **secondary wave**) A *body wave* that causes the rocks along which it passes to move up and down perpendicular to the direction of its own movement. See also *P wave.*

syncline A concave *fold,* the central part of which contains the youngest section of rock. See also *anticline.*

tabulate corals An extinct *order* of corals that lived during the *Paleozoic Era,* characterized by a boxlike structure.

Taconic orogeny Mountain-building event in the northern Appalachians occurring during the *Ordovician Period* resulting from the collision of a collection of continental fragments and island arcs with the North America continent. The Taconic Mountains of New York are the remnants of mountains formed by this collision.

tarn A deep, typically circular lake that forms when a *cirque glacier* melts.

teleosts Group including most modern bony fish, distinguished by symmetrical fins, round scales, and short jaws.

tension *Stress* that stretches or extends rocks, so that they become thinner vertically and longer laterally. Tension may be caused by *divergence* or *rifting.*

tephra (plural noun) Pyroclastic materials that fly from an eurpting volcano through the air before cooling, and range in size from fine dust to massive blocks.

Tertiary Period The first period of the *Cenozoic Era,* lasting from 66 million years ago to 1.6 million years ago. Subdivided into the *Paleocene, Eocene, Oligocene, Miocene,* and *Pliocene* epochs.

Tethys Sea A portion of *Panthalassa* (see glossary definition) that was partially enclosed by the northern and southern arms of the Pangaean continent.

theory A comprehensive explanation of a given set of data that has been repeatedly confirmed by observation and experimentation and has gained general acceptance within the scientific community but has not yet been decisively proven. See also *hypothesis* and *scientific law.*

therapsids Advanced mammal-like reptiles having partially fused skull and jaws and differentiated teeth that lived during the *Permian* and *Triassic.*

thermal expansion A form of *mechanical weathering* in which heat causes a mineral's crystal structure to enlarge.

thermal plume A vertical column of upwelling *mantle* material, 100 to 250 kilometers in diameter, that rises from beneath a continent or ocean and can be perceived at the Earth's surface as a *hot spot.* Thermal plumes carry enough energy to move a plate, and they may be found both at plate boundaries and plate interiors.

theropods Bipedal *saurischian* dinosaurs that were mainly carnivores, exemplified by *Tyrannosaurs rex.*

thin-skinned tectonics Style of tectonic deformation in which platform rocks are strongly folded and faulted while underlying basement rocks are less deformed, typically forming *fold-and-thrust mountains* (see glossary definition).

thrust fault A *reverse fault* marked by a *dip* of 45° or less.

tide 1. The cycle of alternate rising and falling of the surface of an ocean or large lake, caused by the gravitational pull of the Sun and especially Moon in interaction with the Earth's rotation. Tides occur on a regular basis, twice every day on most of the Earth. 2. A single rise or fall within this cycle.

till See *glacial till.*

Tippecanoe sequence The second of the major *cratonic sequences* (see glossary definition) deposited during the *Paleozoic Era,* consisting of *Ordovician* and *Silurian* age *sediments* deposited by *transgression* and *regression* on the platform. The sequence is bounded above and below by *unconformities* in many places.

tombolo A sandy coastal landform that grows from a mainland to a *sea stack,* produced when waves are intercepted by the stack and deposit their sediment load on its landward side.

topography The set of physical features, such as mountains, valleys, and the shapes of landforms, that characterizes a given landscape.

transform motion The movement of two adjacent lithospheric plates in opposite directions along a parallel line at their common edge. Transform motion often causes *earthquakes.*

transgression Rising sea level, causing the inundation of land areas at the continental margins and the shifting of the shoreline toward the interior of the continent.

translatory motion Motion of water in which the water itself actually moves.

transverse dune One of a series of *dunes* having an especially steep slip face and a gentle windward slope and standing perpendicular to the prevailing wind direction and parallel to each other. Transverse dunes typically form in arid and semi-arid regions with plentiful sand, stable wind direction, and scarce vegetation. A transverse dune may be as much as 100 kilometers long, 200 meters high, and 3 kilometers wide.

Triassic Period The earliest period of the *Mesozoic Era,* lasting from 248 million years ago until 205 million years ago.

tributary A *stream* that supplies water and sediment to a larger main stream.

trilobites A group of extinct arthropods that dominated the *Cambrian* oceans.

trunk stream A main *stream* into which *tributaries* carry water and sediment.

tsunami (plural **tsunami**) A vast sea wave caused by the sudden dropping or rising of a section of the sea floor following an *earthquake.* Tsunami may be as much as 30 meters high and 200 kilometers long, may move as fast as 250 kilometers an hour, and may continue to occur for as long as a few days.

unconformity A boundary separating two or more rock layers of markedly different ages, marking a gap in the geologic record.

uniformitarianism The hypothesis that current geologic processes, such as the slow erosion of a coast under the impact of waves, have been occurring in a similar manner throughout the Earth's history and that these processes can account for past geologic